The Aircraft Cabin:
Managing the Human Factor

The Aircraft Cabin: Managing the Human Factors

Mary Edwards and Elwyn Edwards

Gower Technical

Published by
Gower Technical
Gower Publishing Company Limited
Gower House
Croft Road
Aldershot
Hants   GU11 3HR
England

Gower Publishing Company
Old Post Road
Brookfield
Vermont   05036
USA

Electronically typeset by Mary Voake, Brameur Ltd, Aldershot

British Library Cataloguing in Publication Data
Edwards, Mary
    The aircraft cabin : managing the human factors
    1. Aircraft. Operation. Human Factors
    I. Title   II. Edwards, Elwyn
    387.74044

Library of Congress Cataloguing in Publication Data
Edwards, Mary
    The aircraft cabin / Mary Edwards and Elwyn Edwards
        p.   cm.
    1. Aircraft cabins--Human factors. I. Edwards, Elwyn
    II. Title.
TL681.C3E38   1990
629.134'45--dc20   90-30372
                                            CIP

ISBN 0 566 09056 2
ISBN 0 566 09091 0 Pbk

Printed in Great Britain by
Billing & Sons Ltd, Worcester

Contents

Preface

With the exception of Dr. Dan Johnson's excellent book *Just in Case*, little has been done to bring together between the covers of a single book the scattered material pertaining to the Human Factors of the aircraft cabin. Relevant work is to be found in reports, bulletins, journals, and conference proceedings, not all of which are easily obtainable. No doubt much has been overlooked; we would be delighted to hear about omissions.

It is hoped that this book may be of some assistance to each of several groups of readers. Cabin staff, particularly those in supervisory or training positions, may find here an extra dimension to augment their professional knowledge and experience. Similarly, those involved in the design of cabin interiors or equipment might be interested to consider the wider context in which their products are employed. Students of ergonomics (or Human Factors) and allied subjects may find that an account of this one area of application adds some flesh to the topics in their courses. Finally, the great travelling public may welcome a glimpse "behind the scenes" to assist them in understanding the workings of the aircraft cabin. In attempting to write for such a broad readership, we have made compromises in both content and style. Charitable readers will ascribe to this cause all the shortcomings they may discover.

Part I contains an account of the aircraft cabin under "normal" circumstances, whereas Part II is concerned with emergencies. It has, of course, been impossible to maintain rigorously such a division of material. Part II is the longer for the following reason. Much of the Human Factors of cabin design shares a commonality with the general body of knowledge in such areas as human space requirements, seating, noise, and illumination. The relevant material is readily available elsewhere, so that repetition here would be pointless. Aircraft emergencies, on the other hand, present problems of an unusual nature a fact which will be apparent from the overview of such hazards in Chapter 7.

The question of units of measurement is a difficult one. Whilst scientific writing has widely accepted the Système Internationale d'Unités (SI), obvious difficulties arise for readers more familiar with British Imperial units. Degrees of arc and minutes of time, both absent from SI, are helpful to many people. Furthermore, the aviation community is, for good navigational reasons, wedded to nautical miles and knots as measures respectively of distance and speed. In most parts of the world, aircraft altitude is expressed in feet rather than metres. Our strategy is as follows. Most of the data quoted are expressed first in SI units, followed by an approximate equivalent in Imperial or other widely known unit. When formal regulations or quoted research specify values in Imperial measures, the sequence has been reversed. The abbreviations used to express units of measurement are included, along with other abbreviations, in Appendix 1.

No attempt has been made, of course, to present a comprehensive review of legislation relating to the aircraft cabin. On the other hand, decisions promulgated by regulatory authorities are highly relevant to the design and operation of aircraft. Sprinkled throughout the text are examples of regulations drawn either from the UK or US codes. Such regulations, being both summaries of earlier thinking and determinants of future action, are used to give a flavour of the relationship between the technology of Human Factors and the authority of the law. Although British Civil Airworthiness Requirements (BCAR) are being replaced by other codes, references to them have been maintained since they will be familiar to some readers, and many aircraft currently in service have been designed and operated under BCAR. A brief outline describing UK and US codes appears in Appendix 2.

Many people have assisted us in the task of preparing this book. In the early stages we benefited from discussions with our friend the late Dr. Neville Birch, but further collaboration with him was not to be. Captain Frank Hawkins and Ms Suzette Burghard reviewed substantial sections of the draft text and made numerous helpful comments drawing upon their extensive knowledge and experience of airline operations.

Of others who kindly gave of their time and expertise to offer information and advice, we would particularly mention Ms Helen Butler of Britannia Airways; Mr Edward Abbott, Ms Hilary Rudge and Ms Jo Paul of British Airways; Dr Peter Buckle, Mr Geoffrey David, and Mr Ian Randle of the University of Surrey. Miss Kloppie Edwards provided some editorial assistance. It must, however, be clearly understood that we ourselves claim full credit for all the ignorance and misguided thinking which has stubbornly resisted our colleagues' best efforts.

Assistance with illustrations from the following people is gratefully acknowledged. Mr R.A.R Wilson of Historical Aviation Service for Figure 2.1; Haynes Publishing Group for Figure 2.2; United Airlines for Figure 2.3; Mr Hugh Field of British Aerospace for Figures 3.3 and 8.2; Mr Gunnar Ohlsson of SAS for Figure 3.4; British Airways for Figure 5.1; Mr Stuart Haycock of Cossor Electronics for Figure 7.1; Mr J.E. Bellamy of Draeger Limited for Figure 8.5.

Permission to quote from published material was kindly granted by the following: Ian Allan for a passage from *Pictorial History of KLM* by Roy Allen; Captain David Beaty for passages from *The Water Jump* published by Secker & Warburg; Mr. James Brenneman for the table enumerating incidents of active flaming or combustion in the cabin; the Flight Safety Foundation Inc. for passages from Safety Bulletins; The Guardian for a passage from "Slavery at 30,000 feet" by Lindsey McKie; Mr J.M. Ramsden for reproduction from *Flight International* of the table enumerating occurrences of in-flight fires.

Our publisher, Mr John G.R. Hindley, has helped us in many ways.

It has been our intention to respect copyright, and to make all proper acknowledgement of our sources. Apologies are offered in respect of omissions, which will be rectified at the earliest possible opportunity.

M.E. and E.E
March 1990

Tables

Illustrations

PART I
The Components of the Cabin

1   Human Factors?

Effective design and management of the aircraft cabin demands the procurement and use of many resources in order to achieve high standards of safety and service. Human Factors is the technology concerned with people and the ways in which they interact with one another and with the world around them. The relevance of Human Factors to the cabin is introduced by employing the SHEL model which describes systems in terms of three types of resource - Software, Hardware, and Liveware - interacting together and with their Environment. The purpose of this model is to provide a framework in which to organize the issues which arise during the design and management of dynamic real-world systems.

Every passenger travelling by air will be sensitive to the standard of comfort and service offered in the cabin. Experienced travellers are capable of making fine discriminatory judgements between aircraft types, airline policies, and even the performance of the attendants on a particular occasion. Priorities vary; some passengers attach more importance to the amount of space available, others to the quality of food or the provision of such items as newspapers. The appearance and behavioural manner of the cabin staff are unlikely to escape the notice of anyone.

The inexperienced traveller may require more direct personal assistance on entering the aircraft in order to find the allocated seat, stow hand baggage, and fasten the seat belt. Whilst comparative judgements cannot be so readily made, deep and lasting impressions of air travel can result from a few early flights. The standard of facilities available and the quality of service rendered by the crew are usually major factors in the evaluation of the experience.

The importance of providing service to passengers has been recognized within the industry since the introduction of cabin attendants during the 1920s. Analogous services were familiar to passengers travelling by sea. With the growth of popular air transport, operators have become increasingly aware of the impact of cabin service within a competitive commercial environment. Their concern is evidenced by the attention paid to the detailed design of cabin decor, by the type of publicity using illustrations of glamorous attendants providing personal service, and by the attention paid to appearance and deportment during the selection and training of cabin personnel.

Numerous organizations conduct and publish surveys of passengers' evaluation of airlines. Such material provides feed-back to the operators concerning their performance, and indicates the public view of their relative strengths and weaknesses. Furthermore, the publications encourage the travelling public to adopt a more critical and discriminating attitude towards air travel and consequently to demand a continually improving service for the consumer.

Unfortunately, not all journeys by air follow their planned course. A passenger could become ill or sustain injury during the flight; a fire could break out in the cabin; a technical fault in the aircraft might occur, necessitating rapid descent and landing; the aircraft might collide with a bird or with another aircraft such that severe damage is sustained. In all such circumstances it will fall to the cabin crew to take appropriate action in an attempt to ensure the safety of all on board. Clearly, the handling of such emergencies calls for knowledge and skills quite different from those associated with the provision of normal services. In the most extreme cases a complete change of role is required from that of the server to that of a leader and manager whose terse commands should be promptly obeyed by passengers in the interest of survival.

It is with this second set of duties that regulatory authorities are concerned. Before a public transport flight can take place, a prescribed number of qualified cabin attendants must be at their stations. Training courses, checking procedures, and in-flight duties relating to safety are all controlled by national regulatory bodies. Within this framework, activities concerned with public relations and service are left in the hands of each airline, which is free to allocate such additional tasks to the crew.

The major part of the actual work-load of attendants during a typical flight is, of course, devoted to these non-statutory tasks. With the exception of routine cabin checks prior to take-off and to landing, and the presentation of the statutory safety briefings, attendants will devote almost the whole of their energies to providing an efficient service to their passengers. For the majority of attendants, experience in the duties associated with serious emergencies will be confined to the simulated exercises used during the course of training and checking.

The current standards of service and safety provided in the passenger cabin are made possible only by way of a variety of sophisticated technical facilities on board the aircraft, backed up by a complex network of ground-based organizations devoted to the development and provision of the appropriate resources. Many technologies contribute to the availability of food services emanating from small galleys, and to in-flight entertainment programmes, toilet facilities, air-conditioning and pressurization systems, and the items of safety equipment carried in readiness for any emergency which might arise. These achievements in engineering cannot, however, function in isolation. The importance of the human contribution provided by cabin staff has already been emphasised, and to that must be added the activities of persons concerned with the production and maintenance of cabin facilities.

There remains one further contributing factor to consider. Cabin service functions efficiently only if certain rules and procedures are carefully followed. Without such a disciplined approach, chaos would result. In the event of an emergency requiring, for example, a swift evacuation of the cabin, a pre-planned and rehearsed routine is essential in order to ensure an effective performance.

The three-resource system, as described above, typifies enterprises not only within aviation but throughout industrial, military, and even leisure-oriented systems. Almost invariably some equipment, buildings, vehicles, or other type of "Hardware" will be required. In order to ensure that the system functions in an orderly and effective manner, a set of rules, regulations, recognized practices, and standard operating procedures is necessary. Such items are subsumed under the general heading of "Software", and as a means of maintaining similarity in vocabulary the all-important human beings in the system might be called "Liveware". Those whose task it is to construct a system for a particular purpose have available to them these three types of resource - Hardware, Software and Liveware - and it is their task to select and deploy these resources to the greatest advantage.

Little progress can be made in the task of deciding how best to select and deploy the available resources until full account is taken of the Environment in which the system is required to operate. At cruising height, an aircraft operates within extremely hostile physical surroundings. The air is rarefied; the ambient temperature is very low; the atmosphere contains such potentially harmful agents as radiation and gaseous ozone. All such features must be considered as part of the aircraft design process. Unless the proper steps are taken, these hostile features would penetrate into the cabin itself, producing conditions which would fail to sustain the lives of crew and passengers.

The notion of "Environment" may be extended to encompass the social and economic climates in which an airline must operate, and which will exert considerable influence upon its policies. Once again, the necessary steps must be taken to ensure that an airline can continue to survive, and ideally to thrive, within the constraints imposed by these external and unstable conditions.

The description of three types of resource interacting together within the context of their Environment has been called the SHEL Model, a name derived from the initial letters of the component parts (E02). Figure 1.1 illustrates the components and the interactive links. During both the initial design and the on-going management and maintenance of a system, due attention should be paid to its interfaces in order to ensure that all the components are able to work together safely and effectively.

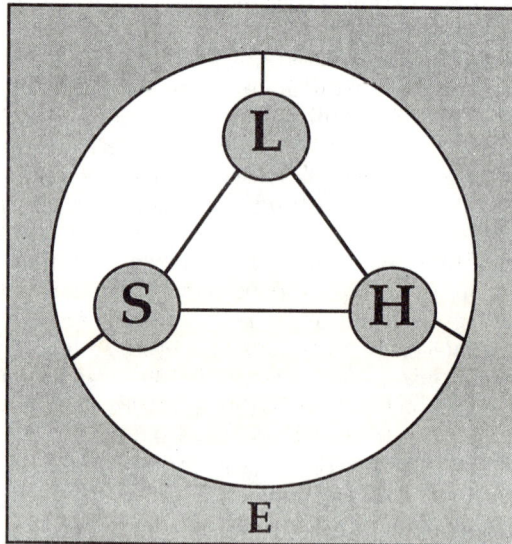

Figure 1.1 The system resources - Liveware, Software and Hardware - interact together and with their Environment.

Human Factors is the name of the technology devoted to the Liveware components in a system. It attempts to ensure that the needs, capabilities, and limitations of human beings are given due account within the processes of system design and management. A few examples might clarify the scope of Human Factors. What is the range of passengers' leg lengths to be accommodated within the pitch of aircraft seats? What level of illumination is necessary for reading printed materials? What is the maximum shelf height usable by a stewardess? How should safety briefing cards be illustrated and worded in order to maximize their effectiveness? How much acceleration can the human body tolerate before permanent damage results? How much torque can a person apply to a door handle? What is the effect on human performance of oxygen deficiency? What are the minimum acceptable dimensions of an overwing escape hatch?

All these questions, and many others, have been investigated, particularly during the last thirty years or so, and the results may be found in the various Human Factors journals, books, and reports. Psychology, anthropometry, and physiology are the foremost branches of science upon which the applied discipline of Human Factors is based. To a large extent, Human Factors bears the same relation to the life sciences as hardware engineering bears to the physical sciences. (The term "Human Engineering" has been employed as a synonym, but has the disadvantage of being confused with exercises in genetic intervention.) Whilst the life sciences provide a wealth of data and techniques, studies carried out within the area of application are essential before any firm conclusions may be drawn.

Initial capital letters have been employed in the term "Human Factors" and there is a special reason for this. The choice of a name for the technology, although long established, has the disadvantage that the words may be used in another sense meaning "aspects relating to the nature of people". Thus it might be said, for example, that about 70% of aircraft accidents are attributed to human factors. Such a statement, whilst true in one sense, illustrates the ambiguity, since it would most certainly be disturbing if any accidents whatever could be attributed to Human Factors!

Similarly, for reasons of clarity, the terms describing the system resources - Software, Hardware, Liveware - together with the Environment in which they operate are printed with capital letters to indicate their particular use within the SHEL model.

The word "Ergonomics" was coined in the UK by K.F.H. Murrell to describe this discipline, and both terms are now used interchangeably. Should a definition be required, we might say that Human Factors (or ergonomics) is the technology concerned to optimize the relationships between people and their activities by the systematic application of the human sciences, integrated within the framework of system engineering (E03). It is, perhaps, easier to convey the essence of such a statement by saying that Human Factors addresses those issues which arise from the inclusion of Liveware within a SHEL system.

When applying this model to the aircraft cabin, it is apparent that the attendants are the Liveware system operators. They must interface together to form an efficient team. The passengers, Liveware system users, also interact with the crew members to form a second type of Liveware-Liveware interaction. Cabin attendants have a range of Hardware items which they must use, and which should therefore have been designed with the characteristics of the users taken into proper account. Similarly the rules and procedures governing cabin operation must fit smoothly both with the limitations of the human users and with the items of Hardware to which they relate. In addition, all of the resources must cope successfully with their Environment.

Such then is the nature of Human Factors and its relevance to the design and management of the aircraft cabin. In the chapters which follow, an attempt is made to summarize some of the relevant Human Factors work which has been carried out and to indicate those areas in which further work remains necessary.

2 The Evolution of Passenger Aviation

Travel by air is a twentieth-century phenomenon and its evolution has been rapid. Some historical landmarks in the development of the passenger cabin and in the introduction of cabin attendants are described here.

In May 1908, less than three years after the epic achievements at Kitty Hawk, Wilbur Wright took aloft the first passenger to experience powered flight. During the period since this rudimentary inception of air transport, the facilities provided for passengers have evolved dramatically. Instead of the single passenger seat in an open cockpit, passengers are carried in pressurized cabins in which a wide range of services are available. The reliability of air transport has improved to the extent that the likelihood of arriving at a planned destination is greater than for most other forms of transport. The number of passengers per aircraft has increased from one to five hundred or more who may be accommodated in a wide-bodied jet.

It must also be recorded that 1908 was the year in which a passenger first met with a fatal accident. Considerations of passenger safety are thus as old as air transport itself.

THE EARLY YEARS

In 1909, there were two aircraft which had been designed specifically to carry a passenger. These were the Wright A, a biplane equipped with skids and launched from a rail, and the Blériot XII, a monoplane with a wheeled undercarriage which held the world speed record of 47.85 mph in that year. In 1911 in California, the first passenger travelled in a seaplane, a Curtiss landplane modified by the designer, which was later fitted with wheels thus becoming the first amphibian. The same Curtiss devised the first seat belt, the "Curtiss" belt, after J.H. Towers, his

9

protégé, had during the early days of flying survived a fall from 1500ft in an out-of-control aircraft. The mandatory requirement for passenger seat belts, however, was to come much later. The first aircraft with enclosed cabins, one a monoplane, the other a biplane, were introduced by Avro in 1912.

A foretaste of the comforts possible in air travel was demonstrated in 1913 by the Sikorski-designed "Bolshoi", the first four-engined aeroplane to fly. Piloted by the designer, this aircraft took off from St Petersburg (now known as Leningrad) in 1913 for a flight lasting nearly two hours with eight passengers on board. The passengers were carried in a fully-glazed cabin which was furnished with four armchairs, a sofa and a table. In the following year, sixteen passengers were carried on a successor of the "Bolshoi" for a flight during which the first airborne meal was served and eaten. However, these events were atypical of aeroplane transport of the period and resembled more the opportunities provided by airships.

The first recorded scheduled airline operated in Florida in 1914, carrying passengers and freight in a two-seater flying-boat the 22 miles between St Petersburg and Tampa. This survived only for four months and further development of civil air transport was curtailed by the First World War. Indeed, air travel in the United States had an uneven history of small, short-lived airlines until the privatization of the US mail transport with its associated subsidies was made possible by legislation in 1925.

POST WORLD WAR ONE

After the First World War, the requirement for the transport of personnel and documents from London to Paris in connection with the work of the Peace Conference during late 1918 and 1919 gave a fillip to European air travel. This period saw the initiation (and subsequent financial failure) of domestic air services in Germany, France and Great Britain. In 1919, international airlines were founded in France and Holland. The Royal Dutch Airline (KLM) survives to this day. Although Britain's first international airline, Aircraft Transport and Travel (AT&T), was founded in 1916, it did not operate until 1919, when in August a daily service was introduced between London and Paris. In September, Handley Page Transport inaugurated its services between London and Brussels, and between London and Paris. However, AT&T went out of business at the end of 1920 and Handley Page Transport ceased their cross-channel operations early in 1921.

With the financial failure of the unsubsidized British companies came the recognition that government assistance was essential to support the air transport industry. This led to the formation in

1924 of Imperial Airways by the combination of four air transport companies, echoing the Railways Act of 1921 which provided for the amalgamation of a number of smaller railway companies to form the four major operators in 1923.

The aircraft used in early post-war years for transporting passengers were adapted from war-time bombers. Passengers flying in open cockpits were provided with leather overcoats, helmets and hot water bottles for protection against the elements and cotton wool to shield their ears from the high noise levels to which they were exposed.   Not all enclosed cabins had windows.   The Handley Page O/400, for example, was converted from a war-plane and before subsequent modification carried up to a   dozen passengers early in 1919 in considerable discomfort, in a noisy and draughty fuselage.

A more successful commercial career was enjoyed by the Vickers Vimy, a bomber for which huge orders had been placed too late for the aircraft's operational use prior to the armistice.   This two-engined machine, which performed the first non-stop transatlantic crossing in 1919, carried up to fifteen passengers in its fully enclosed cabin.   An interesting item of equipment   was a telephone for passengers to communicate with the pilot who was located in an

Figure 2.1 Cabin of the Armstrong Whitworth "Argosy", a three-engined biplane of which seven were built for Imperial Airways.

open cockpit which could also accommodate an additional passenger.
Comfort was not one of the main attractions of flight in the

early 1920s.   Indeed, the experience could be extremely unpleasant with low temperatures, high noise levels, noxious fumes and excessive vibration.   Sickness in passengers was not uncommon, and buckets were provided for their convenience.

A notice to passengers travelling with KLM in the early 1920s advised them to "keep away from the propeller and tell your fellow passengers to do the same.   Don't hang over the side of the aircraft or bother your pilot;   don't stick your arms or head suddenly overboard - you could lose your hat.   Don't throw things overboard; refrain from alcohol several hours before a flight, and don't eat split peas, beans, brown bread or any food which could cause excessive gas in the intestines.   Visit the toilet before leaving." (A04)

Rapid developments were taking place, however, in the design and furnishing of the passenger cabin.   Open cockpits gave way to enclosed, glazed passenger cabins furnished with wicker chairs, chosen for their light weight.   Wicker chairs were later superseded by fixed upholstered seats and the interior of the cabin grew reminiscent of a railway carriage with fixed seats in rows, and mesh overhead luggage racks.

Cabin development was obviously dependent upon advances in the technologies of aircraft structures and propulsion.   Whilst the victorious allies of World War I were attending to the conversion of bombers, work in Germany proceeded in contravention of the Versailles Treaty upon the development of all-metal passenger aircraft.   The first of these, the Junkers F13, remained in passenger service between 1919 and 1940 at which time it was transferred to the Luftwaffe.   Other Junkers aircraft included the W33, which made the first east to west Atlantic crossing, the diesel-engined Ju 86 and the three-engined Ju52-3m, one of the most successful transport aircraft of all times.   (See Figure 2.2)

The Fokker company, which also produced a range of transport aircraft during the inter-war period, pioneered numerous innovations. The FVII series (the best-known example of which was the "Southern Cross") introduced the first on-board lavatory   in 1924.   This aircraft also had sliding windows which could be opened in conditions of turbulence to provide fresh air and thus, it was claimed, reduce the incidence of airsickness. During the following year the first in-flight film was shown, a luxury which failed to become established for many years.

In 1926 Imperial Airways took delivery of its first Armstrong Whitworth Argosy, a three-engined biplane carrying twenty passengers, ten on each side of the cabin.   This machine was employed upon the "Silver Wing" service between London and Paris during which drinks and lunch were served. Another tri-motor biplane, the de Havilland Hercules, was used for the longer Empire routes   and   was   later joined by   the   four-engined   Hannibal   and

Figure 2.2 Cabin of the legendary Junkers Ju52-3m.
Passengers were provided with spacious adjustable seats.   Only coats
and hats required accommodation in the cabin, other items being
stored in the rear compartment.   The windows were fitted with
roller blinds.   The lavatory door can be seen in the rear bulkhead
on which an altimeter is fitted.

Heracles machines.
   As the 1920s drew to a close, America took the first steps
towards establishing its pre-eminence in the production of airliners,
a position it has retained to the present day.   The first aircraft in
a long line of successful types was the Ford 4-AT, which carried
ten passengers and which, incidentally, achieved in 1929 the first
flight over the South Pole.

THE NINETEEN-THIRTIES

By the 1930s, a high standard of comfort was available to
passengers both in flying-boats and in the bigger land planes.
With their large hulls, flying-boats typically resembled ocean liners
in their facilities, with sleeping cabins, promenade decks and dining
rooms, and they were used extensively by Pan American World
Airways in its pioneering scheduled flights across the Pacific.   The
Sikorski S-40 American Clipper of 1931, inspired by the Cunard
liners and capable of carrying up to 50 passengers, was luxuriously
fitted with carpets, walnut walls and ceilings, upholstered Queen

Anne chairs and, extending the nautical flavour, life belts hanging from the walls of the lounge. Hot meals, prepared in the aircraft, were served at tables laid with linen cloths and heavy silverware. Sleeping compartments were available, which were shared by taking turns.

Perhaps one of the most luxurious aircraft of the pre-World War II period was the Boeing B-314, a flying-boat which flew first in 1939 and which could seat up to 74 passengers or sleep 40 in berths. It had five passenger compartments including a honeymoon suite, a dining room seating fifteen which provided waiter service for the seven-course meals accompanied by fine wines, and separate toilet compartments for men and women. Passengers could stroll about or play bridge during the flight.

The Short Brothers Empire flying-boats which first flew in 1936 were also luxuriously equipped. These aircraft continued to fly the line with British Overseas Airways Corporation (BOAC) until 1947.

For the immense distances involved in travelling throughout the British Empire, Imperial Airways utilized both flying-boats and land planes. The Handley Page HP42E Heracles was a four-engined biplane which provided a high standard of comfort for 38 passengers, with sound-proofing, heating and ventilation. The dream of a regular passenger service to Australia was achieved in 1935. The journey, extending over 13,000 miles, took twelve days to complete.

The four-engined Junkers G38 monoplane, carrying 34 passengers between London and Berlin, had a smoking saloon in the rear of the aircraft.

AIRSHIPS

The appearance of the airship on the aviation scene must not be overlooked. The first of its kind to fly was a Zeppelin in 1900, and 1910 saw the start of a regular passenger service between various German cities. In the years 1909 to 1914, seven airships transported more than 34,000 passengers on more than 1500 journeys without a single fatality or injury. Indeed, until the Hindenburg disaster in 1937, not one fare-paying Zeppelin passenger had been lost (though a number of serious accidents had happened to these craft) during nearly forty years.

Travel by airship was very similar to travel in a luxury ocean liner. Even at a time when aeroplane passengers were huddled in open cockpits, immobilised for the period of their journey, airship passengers were accommodated in long saloons seating up to 24 passengers in comfortable chairs, with meals served by an attendant steward. Later airships such as the Graf Zeppelin and the Hindenburg, the latter carrying fifty passengers, were constructed with saloons (the Hindenburg carried a grand piano), individual passenger cabins, bathrooms, lavatories, and dining rooms provided with heavy linen, silver, and even flowers for the tables.   Catering staff cooked elaborate meals in specially equipped kitchens.   These craft flew at speeds of less than 100 mph and at relatively low altitudes, allowing passengers to enjoy detailed aerial views of the terrain.

The early successes of the airships, including the return journey in 1919 of the British R34 to the United States, was followed in 1921 by the crash of the R38.   This airship, which had been sold to America, crashed into the River Humber on a test flight, killing more than forty people.   Later in the 1920s, the British government was persuaded to fund the construction of two airships, the R100 and the R101.   They made their first flights in 1929. However, after a tragic accident to one of these in which 48 out of 54 of those on board were killed at Beauvais in 1930, the second was sold for scrap and further development in Britain was halted.

In 1929, a Zeppelin flew around the world in four stages, completing the journey within twenty-one days.   A regular service from Germany to South America was inaugurated in 1931 and this was followed later by a regular service across the North Atlantic.

Perhaps the most well-known airship was the Hindenburg.   This, the largest and most luxurious of the type, first flew across the Atlantic in April 1936.   Its last flight was in May 1937 when it inexplicably burst into flames as it was mooring at Lakehurst, New Jersey, in the United States, killing one-third of those on board. This accident, vividly and unforgettably described by radio commentator Herbert Morrison as it took place, spelled the end of an era.

THE "MODERN" AIRLINER

It is generally agreed that the United States was the birthplace in the 1930s of the modern airliner. During the previous decade, Europe had maintained leadership in air travel and it was not until legislation was enacted in 1925 to enable the United States Post Office to put its airmail service to private tender that air passenger transport became viable.   This was because of the subsidy paid by the Post Office for the carriage of mail which enabled the airlines to carry passengers at a profit. The early concentration of the United States air services on the carriage of mail had significant consequences for the lead taken by that country in transport aviation.   In order to ensure prompt delivery of the mail and to build a reputation for punctuality and reliability, it was necessary to fly in bad weather and during the hours of darkness. Without the constraint of safeguarding passengers, it was feasible to attempt this.   Both the experience of flying in a wide range of weather conditions and the development of the navigational aids which assisted the pilots in these conditions were to be of considerable benefit to passenger transport during this period.

Three aircraft are associated with the beginning of the American domination of aircraft design and production.   These are the Boeing 247, the Douglas DC-2, and the Lockheed Electra.

In 1933, Boeing introduced the 247, a low-wing two-engined aircraft carrying twelve passengers.   Because the Boeing Company, being heavily committed to United Airlines, was unable to satisfy the demands of all its potential customers for this aircraft, the Douglas Company developed from its DC-1 prototype a number of the DC-2 aircraft for Trans World Airlines.   The Douglas aircraft had the advantage over the Boeing in that the struts did not cause an obstruction by passing through the cabin.   The DC-1, with its upholstered seats, a forty-inch seat pitch, foot-rests, and individual reading lamps, incorporated the latest developments in sound-proofing, heating, and ventilation.   Only one aircraft was built and it was succeeded in 1934 by the DC-2 which carried its fourteen passengers in a sound-proofed, ventilated cabin, with curtained windows, individual air vents, and single fixed, but adjustable, seats. The first European example of this type was supplied to KLM and gained fame in the 1934 air race to Australia.

Because American Airlines wanted to fly a transcontinental sleeper and the DC-2 was too narrow to accommodate sleeping berths, the DC-3 was developed by scaling up the DC-2 with a wider fuselage and increased length and wing-span.   By the late 1930s, most American passengers were being carried by the 21-seat DC-3 "Dakota", one of the most successful commercial transport aircraft.

Less epoch-making but worthy of note was the Lockheed L-10 Electra which made its appearance in 1934.   Although smaller than its rivals, this aircraft incorporated many of the same innovative design characteristics as the Boeing and the Douglas and, like them, was the first in a series of major airliners produced by the company.

Further developments in aircraft design led to the Boeing 307 "Stratoliner" which made a major contribution to passenger comfort in 1940 with the first pressurised cabin to go into airline service.

POST WORLD WAR TWO

Although civil passenger transport was almost halted during the Second World War, enormous technical developments in aircraft design had taken place during the years of conflict.   Immediately after the war, the emphasis was largely on luxury travel in the air in an attempt to encourage passengers to fly rather than to travel by land or sea. Flying-boats were still in use until the early 1950s but the leisured luxury associated with this method of travel was not destined to survive. Their advantages, deriving from the shortage of aerodromes, the comfort of the accommodation, and the confidence they inspired in passengers over large expanses of water, declined as a consequence of the development of convenient inland airfields. Prominent amongst the post-war aircraft were the Douglas DC-4 Skymaster, the pressurised DC-6, and the Lockheed Constellation. The Boeing Stratocruiser of 1947, developed from the Superfortress bombers, could echo the opulence of the flying-boats. One version provided space to walk around the double-decker passenger cabin which was equipped with berths, a bar, and five attendants for a maximum of 47 passengers who were served seven-course meals en route. However, the appeal to the top end of the market of luxury comparable with that provided by ocean liners was gradually being replaced in the airline industry by the recognition of the economic advantages of cheap mass travel. The shape of things to come was heralded by the advent of the jet-engined airliner and the agreement of the members of the International Air Transport Association (IATA) to the introduction of Tourist Class fares in 1952.

The age of mass transport had arrived, though this was not yet obvious when the de Havilland Comet took to the air with seating for 36 passengers. Jet aircraft flying high above the weather provided a smoother ride for passengers, thereby almost eliminating the incidence of airsickness, with reduced vibration and noise levels. The economics of flight favoured cheap fares and increased throughput.

From the 1950s through to the 1970s, passenger aircraft became bigger. The Boeing 707 which made its inaugural transatlantic flight in 1958 could seat 84 passengers and the Douglas DC-8 made its debut in 1960 able to carry between 116 and 179 passengers. In 1965, the stretched DC-8 was able to carry up to 259 passengers and in 1970, Pan American inaugurated the "Jumbo" Boeing 747 service between London and New York with capacity for 490 passengers.

The introduction of supersonic flight with Concorde in 1976 gave a new dimension to passenger aviation, transporting 128 passengers at 1350 miles/hr. However, because the fuselage is so narrow, there are space restrictions in the cabin and the glamour of Concorde derives from its speed, the exclusiveness associated with high fares, and the special treatment accorded to the passengers on the ground. It also permits those with pressing business to achieve the ultimate status symbol of a day trip from London to New York and back.

THE NINETEEN-EIGHTIES

Whilst the 1970s gave rise to the appearance in airline services of the gigantic Boeing 747 and of the Anglo-French Concorde, no such dramatic innovations can be associated with the 1980s. This is not to say, of course, that the aircraft industry became dormant, but rather that developments during the decade were of a different character.

On the flight-deck, change was more apparent. The "glass cockpit", comprising electronic rather than electro-mechanical instruments, became established, and advanced "fly-by-wire" control systems were introduced. Sophisticated flight management systems provided high levels of support in the tasks of navigation and efficient use of power. Engines became quieter, more economic in the consumption of fuel, and produced less atmospheric pollution.

As a consequence of automation, the size of the flight-deck crew in many aircraft was reduced to two. The progressive disappearance of the flight engineer was regretted in many quarters. Traditionally, the engineer provided the first-line assistance with technical problems in the galley or other parts of the cabin.

Various regulatory issues affected the aircraft cabin during the 1980s. The decision of the US government in 1978 to introduce deregulation into the airline industry had major repercussions. There was a re-grouping of airlines under different ownership, with some established companies disappearing as the result of financial failure. Competition between airlines had become intense, with consequent cost-cutting to remain competitive. The consequences of deregulation for cabin staff was in many cases a severe reduction in pay and related benefits such as pensions. Passengers benefited from lower fares, but these were attained at a cost of reduced punctuality and service.

Society has become less tolerant of smoking in recent years and prohibitions have been implemented in numerous transport systems and in other public places. The US introduced the first air transport ban, applicable on the shorter domestic routes. Australia and Canada followed with total bans applying to domestic operations.

The question of "carry-on" baggage remains under review. The obvious advantages to passengers in avoiding the risk of loss and the delays at destination airports is set against the hazard of items in the cabin causing injuries by falling from overhead luggage bins or cluttering escape routes in the event of an emergency evacuation.

Accidents during the 1980s caused the question of smoke hoods to be reconsidered. The use of water-mist within the cabin provides an alternative approach to protection from fires. Research results have indicated that comparatively low flow-rates of water can confer very substantial benefits (B11).

Figure 2.3 This United Airlines Boeing 747 exemplifies the spaciousness of a contemporary airliner.

Perhaps the most significant issue of the late 1980s illustrated the way in which the industry became a victim of its own history. The problem, commonly referred to as that of "geriatric jets", concerns the number of jet aircraft which have exceeded their economic design lives. Within the Boeing fleets world-wide by the end of 1988, all the 720s, about 60% of the 707s, and 25% of the 727s were beyond their twenty-year economic design life-span. In 1990, the 747s began to reach this age, but more than 10% of them had, at the beginning of 1989, exceeded the economic design figures in respect of hours flown. These data serve as a tribute to the quality of the aircraft, particularly when account is taken of the high prices obtainable on the used aircraft market for the fleets from operators whose maintenance records are highly respected. However, the accident record demonstrates some of the risks involved in the operation of older aircraft, and manufacturers, certification authorities, and airlines have devoted their attention to establishing standards of maintenance which are essential in order that the older aircraft can continue to provide service with safety.

3 The Cabin

> The cabin must cater for the various needs of passengers during the course of their journey. The same space serves as the work place for the crew whose needs must also be satisfied. The furnishings and fittings, levels of lighting and of noise, air quality and radiation risks, together with the standard operating procedures, all fall within the province of Human Factors.

REQUIREMENTS

To the travelling passenger, the cabin serves as a sitting room with entertainment included, a dining room, and sometimes a bedroom. It must also provide washroom facilities and a limited amount of storage space for personal belongings. Passengers' evaluations of the cabin and, to a large extent, of the airline will depend on the extent to which these various needs are satisfied. There are, however, alternative viewpoints on the assessment of this section of the aircraft. To the cabin crew it is a workplace in which certain duties must be performed, often within tight constraints of time and space. To the airline executives, the aircraft fuselage - of which the cabin forms the principal part - is a payload bay: it is the *"raison d'être"* of the aircraft and indeed of the airline. Flight crews, by contrast, have been heard to describe the fuselage as the tube which holds on the tail!

Much as passengers might wish for spacious and comfortable accommodation, and cabin crews might desire more generous working dimensions, designers will be acutely aware that the aircraft must be sold in a highly competitive commercial environment in which purchasers will have taken a hard look at seat costs per passenger mile, and numerous other economic indices. Further constraints will have been imposed by the necessity to satisfy the certification authorities' requirements regarding the integrity and safety of the aircraft.

The capacity of the aircraft and its operating range will have been determined by the manufacturer, whose analysis of market requirements will have led to decisions concerning economic viability. Potential customers may well have been influential at this early stage. Thereafter, the detailed shape of the fuselage will be determined by considerations of aerodynamics and structural requirements and within this fuselage the cabin, along with cargo holds and other items, must be accommodated.

It is customary for passenger doors to be located on the left side of the aircraft and for cargo and service doors to be located on the right side. This separation allows galleys to be serviced and baggage to be handled without conflict with passenger embarkation and disembarkation. The cabin doors on the right side, which also serve as emergency exits, frequently lead directly into a galley area.

The basic cabin offered to airline customers may comprise no more than the empty space, with doors in a given location and with seat rails on the floor. Thereafter customers may be allowed a wide range of options in the arrangement of internal partitions, seat pitch and width, location of toilet compartments and details of interior finish. Space must be made available for certain items of safety equipment required by regulation.

Every facet of the cabin is thus the result of compromises between the various aspects of its construction and use, and the ergonomics considerations must be set within this context.

BASIC FEATURES

In addition to the space allocated to each seated passenger, provision must be made for galleys, toilet compartments, and of course, for aisles. The comfort of passengers, the requirements of the cabin crew, together with considerations of such emergencies as sudden evacuation in the event of a fire on the ground, all contribute to the design of the cabin area.

Head clearance
Because of the shape of the fuselage, maximum headroom is available when the aisle is in the centre of the aircraft. However, for some tall passengers this is insufficient for adequate head clearance and they have to stoop as they move along the aisles. The position of the overhead lockers reduces head clearance and some passengers seated below them have to stoop as they leave their seats.

When galleys are located at the sides of the aircraft, or in upper or lower decks, the ceiling height may be lower than the stature of tall cabin staff who must consequently carry out their tasks in a stooping posture.

Aisle width
Minimum dimensions are defined by regulation. In the UK, for example, the main aisle must be not less than 15in (381mm) wide to a height of 25in (635mm) above the floor and 20in (508mm) wide above that point (BCAR D4-3 4.2.5).

The advantages of wide aisles are that they allow passengers ease of access to their seats, particularly when they are carrying their hand baggage. They also permit passengers more easily to pass each other and to pass service trolleys which can cause a blockage of traffic when in use.

Like other facets of space allocation, the width of aisles is related to class of accommodation with the widest aisles found in First Class and the narrowest in one-class aircraft configured for low cost tourist flights.

The floor
The aircraft floor has been identified as a source of problems for flight attendants because of the slope. The floors in some aircraft are at a slope of about 3° during cruise and this creates extra demands on flight attendants as they walk about the aircraft. Greater effort is required when manipulating service carts up or down an incline compared with pushing along a level surface. In addition, during climb and descent the floor will slope even in an aircraft which is otherwise level. This latter problem is likely to be more acute in short-haul flights when a smaller proportion of the flight-time is spent at cruising level.

Lockers
The overhead lockers in the passenger cabin have evolved from the mesh nets, similar to those in railway carriages of the same period, found in early passenger cabins and intended mainly for coats and hats. Enclosed lockers are more compatible with the developments in the interior styling of the passenger cabin and they allow items to be stored more conveniently and safely.

The height of lockers can be problematic, however, both to cabin staff and to passengers. In some aircraft, shorter individuals will have difficulties in reaching up to open a closed locker and even more difficulty reaching up to close an opened locker.

For a number of reasons, lockers are currently used to carry increasing numbers of items much heavier than hats and coats. Both holiday-makers and business executives find it convenient to bring on board a considerable amount of hand baggage. Returning holiday-makers bring souvenirs for which they have no room in their hold luggage. They also acquire tax-free alcohol after they have surrendered their hold baggage. Business executives wish to reduce to a minimum the time spent waiting for their hold luggage to be brought off the aircraft, and so favour carrying all their belongings into the cabin with them. These developments are not always discouraged by airlines, for which there may be economic advantages deriving from reduced baggage-handling costs and increased availability of space for revenue cargo.

Although FAA regulations (FAR 121.589) require that lockers be placarded with the maximum weight they can contain, it is not always easy to estimate the weight of items stowed in lockers.

PASSENGER FACILITIES

Seats
For the passenger, the seat is the most important aspect of cabin furnishing as this is where most of the journey time will be spent.

The design of passenger seats is the result of a compromise between the requirements of the regulatory authorities which specify the strength; of the operators whose need is for low weight, structural strength, and easy, low-cost maintenance; and of the passengers whose need is for comfort and ease of ingress and egress.

Cabins may be configured in a variety of ways resulting in different densities of seating. The amount of space allocated to a First Class passenger on a long-haul flight, for example, may be more than four times that allocated to a passenger on a short-haul holiday charter.

Much more than floor area occupancy is involved in the provision of good seating. The function of any type of chair is to provide the body with appropriate support in relation to the activities to be performed. The function of an aircraft seat closely resembles that of the domestic armchair. Ideally, it should provide a range of adjustments to suit different sitting postures: upright for eating or writing, reclined for relaxation and sleep. Obviously, the space penalties of an adjustable seat make this an expensive luxury in an aircraft.

Seat comfort is achieved only by taking into account certain basic requirements deriving from human anatomy. Most of the weight of the body should be supported around the two areas of the buttocks beneath the ischial tuberosities, the downward projecting sections of the pelvis. The remaining weight is distributed principally between support in the small of the back, and limb support by way of arm-rests and having the feet on the floor.

Two critical dimensions of the seat pan are its height and depth. Ironically, it is the short person rather than the tall one who is most likely to be disadvantaged by ill-chosen seat heights and depths. Very little weight should be supported by the underside of the thigh, and there should be no pressure applied to the back of the knee, otherwise the blood supply to the lower leg is restricted. The ideal solution comprises seats having adjustable heights. The cost, space, and weight of such a mechanism are deterrents for normal passenger seating, although made available on pilots' seats since it is important that the pilot's eyes be accurately positioned to achieve adequate visibility. Consequently, seat heights should be chosen taking into account the smaller end of the population range. A height of about 380mm (15in) for the leading edge of the seat is probably the best compromise for a general purpose armchair. This is about 50mm (2in) lower than the height recommended for office chairs. The seat should slope 20° - 25° from the horizontal. Seat depth of about 432mm (17in) is appropriate for this type of chair.

The most common error in the design of back-rests is the provision of the major support in the region of the shoulder blades. This is far too high. The principal support should be in the lumbar region (the "small" of the back) thus allowing a curvature of the spine which minimises pressure between adjacent vertebrae.

A head-rest is obviously necessary for relaxing and sleeping. Arm-rests should be about 200mm (7in) above seat level.

Seat width and pitch, the horizontal distance between similar points on seats in successive rows, should be selected to suit people at the larger end of the population range. The width of the seat surface should be not less than 400mm (14in). For side-by-side seating, additional width is required to accommodate the arms. Passengers will be familiar with the problem of attempting to eat in closely spaced seats with a narrow shared arm-rest. For reasonable comfort, the total width for each passenger should be not less than about 560mm (22in). Advertisements for one US airline showed illustrations with accompanying dimensions to contrast its Business Class seat width of 24.25in (616mm) with those of its competitors which ranged from 19in (480mm) to 21.5in (545mm).

Economy Class seating might utilise a 710mm (28in) pitch whereas twice this dimension might be available in First Class seating. There are three penalties of low pitch values. First, access to seats other than those adjacent to an aisle is difficult. Even under normal conditions many passengers, particularly the elderly, will encounter problems. In an emergency, evacuation times may be increased. Secondly, tall people will experience discomfort due to the problem of accommodating the thigh. Thirdly, there is no possibility of altering the angle of the back-rest, and variations in posture are thereby restricted.

Seat strength requirements are based on the inertial loads assuming that the occupant weighs 170lb (77kg) (see page 133). The regulatory authority recognizes that additional consideration should be given to the loads imposed during general usage, particularly when a heavy occupant uses the top of the back of the seat in front as an aid to rising out of the seat (BCAR D4-4 Appendix 1.1.3).

Seat cushions inevitably become compressed and distorted with constant use with the result, in extreme cases, that the occupant is in direct contact with the seat structure. These cushions should be inspected and replaced as necessary or seat comfort will be considerably reduced. Arm-rests and seat recline mechanisms are also liable to damage as a result of normal wear-and-tear and require regular maintenance.

The introduction of movable arm-rests on some aisle seats allows an easier transfer of disabled passengers from in-flight wheelchairs to the aircraft seats.

Seat belts and harnesses
Lap belts are used on forward-facing seats. Those seats which face rearward are also equipped with shoulder harnesses which are necessary for torso restraint during take-off. Extension belts are available for passengers who, on account of obesity or pregnancy, are unable to use the standard belt. Belts attaching to the fitted seat belt are also available to secure infants who are carried on the laps of passengers.

Entertainment
During the 1920s, a wireless set was occasionally provided on board, and in 1925 the first airborne film show took place. It was not, however, until the advent of wide-bodied jet aircraft that on-board visual entertainment became commonplace. Developments in display technology later made possible the use of individual screens located in front of each passenger, replacing the earlier optical equipment. The "third generation" of visual entertainment provides a wide range of options including films and other programme material, games and puzzles, and various types of information. A choice of language is available. Such a computer-supported system, being interactive, allows the passenger to order tax-free goods and other items from the selections offered. Considerable possibilities exist for using the system as a means of presenting safety briefings and other information relating to the flight.

Human Factors considerations include the factors affecting the visibility of the screen together with the safety aspects consequent upon its position.

Toilet compartment
Constraints on space are evident in the design of these compartments and both obese and disabled people are likely to experience difficulties when using them, both in gaining access and in moving inside the compartment. For some disabled passengers, the use of the lavatory may be impossible.

The provision of lavatories, in common with other facilities, reflects the class of accommodation. For example, in one configuration of a wide-bodied aircraft, there are two lavatories for 28 First Class passengers and five for the remaining 244 passengers.

The doorway is typically less than 610mm (24in) wide and there is little space within the compartment to move about. The range of paper goods and the different receptacles for waste may be confusing to inexperienced travellers. Spring-loaded apertures may be too heavy for infirm travellers to operate.

Regulations provide that the doors of toilet compartments be designed to preclude the possibility of occupants becoming trapped inside the compartment, and any locks must be capable of being operated from the outside without the aid of a special tool (FAR 25.783).

The use of the lavatory may be impossible for some handicapped passengers due to the lack of space in which to manoeuvre an in-flight wheelchair outside the compartment as well as difficulties in moving about inside in order to transfer from the chair to the lavatory seat. In-flight chairs with fold-down backs reduce the difficulties to some extent once the passenger has entered the compartment.

Inside the toilet compartment, disabled passengers require certain special features. These include grab-handles and a call button, together with appropriate knobs, taps and other controls all within easy reach.

SERVICE FACILITIES

The galley
Reference has already been made to the problems of low ceiling heights in some galleys and the consequent low headroom for taller cabin staff. The curvature of the fuselage of the aircraft results in ceiling heights being lower in parts of the galley furthest from the centre line.

Space is at a premium in the galley and the most use is made of all that is available. This often results in heavy items being stowed in high places. These may be difficult for shorter cabin staff to reach and also for them to lift down. Less strength is available at arm's length than at positions closer to the body.

The galley is the most dangerous area of the cabin for flight attendants both in normal flight and particularly in periods of turbulence. Just as the domestic kitchen is the major source of injury in the home, so the galley is the area where most injuries to flight attendants take place. Cuts from knives, scalds from hot beverages, burns from fires produced by burning fat or from electrical faults are some of the injuries sustained in the galley.

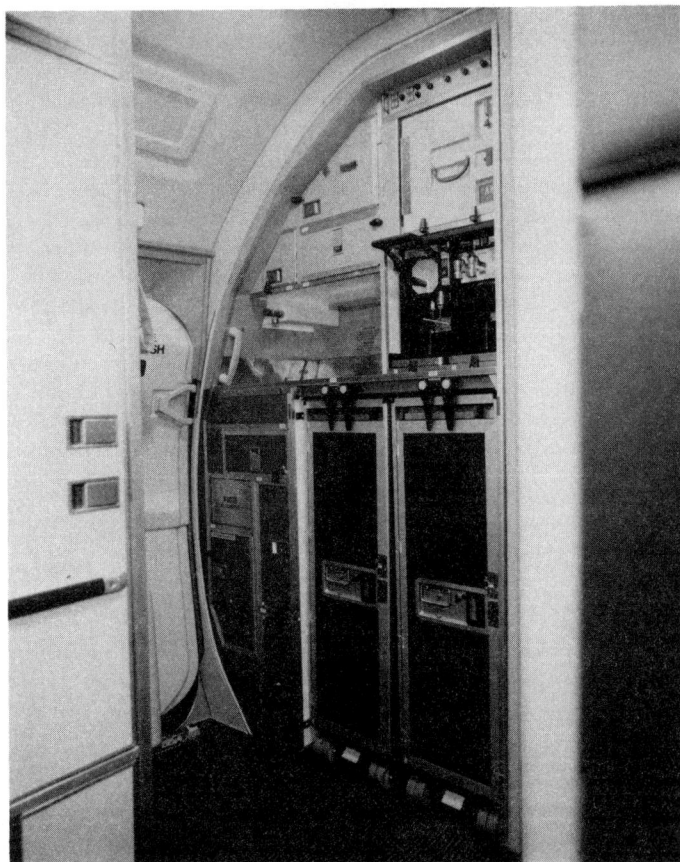

Figure 3.1 A galley in the BAe 146. Due to the necessity of stowing service carts, the height of the work-surface above the floor is about 180mm greater than the preferred value. The service door serves as a Type I emergency exit.

During conditions of turbulence, flight attendants are particularly vulnerable to injury from unsecured items such as knives, hot coffee pots, vessels needed to boil water, heavy articles falling from cupboards and even badly fitted ovens falling from their attachments.

The galley serves as a "common room" or an office for cabin staff who are required to carry out some clerical duties and who also need to have an area in which to remain when the passengers might be sleeping or watching a film and thus require no immediate attention. Sleepless passengers tend to congregate there when the cabin is in darkness.

Service carts

Service carts are frequently stowed under work surfaces. As these carts may be 1.09m (43in) in height, the corresponding height of the work surface is considerably greater than would be recommended for most of the population. Work carried out at this height is therefore both tiring and awkward, leading to increased risk of accidents particularly where the handling of hot liquids is involved.

Carts have given cause for concern on account of the injuries with which they have been associated. In a questionnaire administered to Scandinavian Airlines System (SAS) cabin staff, 85% of respondents found their work heavy because of unsuitably designed equipment. In answer to an open question, more than 20% of respondents regarded the food and drink carts as creating the most serious problems on board an aircraft (W06). These carts can weigh more than 110kg (240lb) when fully loaded and have been implicated in the development of repetitive strain injury, back strain and arm strain. Cabin crew have sustained burns from the spillage of hot liquids such as tea and coffee. In addition, serious injuries have resulted in conditions of turbulence when unsecured service carts become dangerous projectiles in the cabin.

The shape of service carts is dictated by the requirement to fit into narrow aisles. This leads to certain difficulties in their use, including loading and unloading. The narrow working space provided by the top surface can make it hazardous to pour hot coffee or to move items about. Their shape renders them unstable and liable to topple.

Because the service of meals or the sale of goods can be time-consuming and causes blockage of the aisles, there is a tendency to load the carts to the maximum to provide a wide range of items and to obviate the need for extra journeys to re-load. The heavy weight, the small wheels with which the carts are equipped, and the design of the handles with which to push the cart all contribute to the strain on the back and arms. This may be compounded by the slope of the floor.

The development of repetitive strain injury has been attributed in part to the design of the hand brake mechanism of service carts. Carts should be equipped with a braking mechanism which is effective, easy to use without causing strain from repeated movements, and readily operable from either end of the cart. Some carts are fitted with foot-operated brakes. A well-designed braking mechanism will encourage the use of the brakes during service.

Good practice is important in relation to securing service carts both during the time they are in use and when they are stowed away in the galley. In order to immobilize carts during take-off and landing, and also during turbulence, a tie-down mechanism is usually provided in the galley. These tie-down points may also be found in the cabin and clearly it is important that the number of carts in use in the cabin is not greater than the number of tie-down points available. Carts left unattended should always be firmly secured so that they do not become hazards during turbulence.

Service carts containing drawers or cupboards should be equipped with secure locks such that the contents are not spilled during turbulence or in the event of an emergency.

Good practice in the way that the carts are loaded will have an impact on health and safety. Overloading leads to back and arm injuries. Carts in one airline were found to be routinely loaded with up to 45kg (100lb) more than the approved weight (M27). However, it may not always be clear what constitutes the designed weight limit in relation to the variety of goods which are to be moved within the cabin and some guidance may be required. In addition to overloading, problems may arise from goods being loaded in a hazardous manner. To ensure stability, heavy items should be loaded near the bottom and lighter items at the top. When loading hot beverage containers, consideration should be given to their position and to the way they will be used. Thus, spillages and burns from hot liquid may be avoided.

Figure 3.2 The coffee-pot designed by Ergonomi Design Gruppen AB for Scandinavian Airlines System

A well-designed coffee pot
Heavy coffee and tea pots which put stress on hands and arms and which may lead to burns and scalds are a common feature of aircraft catering.   A coffee pot has been produced for Scandinavian Airlines System (SAS) which has been designed to make a significant contribution to the reduction of these problems.   This pot, fabricated from strong, light-weight, insulating material, incorporates a low centre of gravity to provide stability.   The long, dripless spout is designed to allow the pot to be emptied without the necessity of bending the wrist;   the insulated handle is curved and fitted with ribbing to allow a firm grip without putting excessive strain on the hand and arm. This utensil, illustrated in Figure 3.2, serves as a good example of the benefits derived from the application of HF   to the design of Hardware.

Maintenance
In view of the complexity of large transport aircraft, it is almost impossible for every item of equipment on board to be in perfect working order.   A Minimum Equipment List (MEL) is contained within the Flight Manual, the document which requires the approval of the regulatory authority.   The MEL defines the deficiencies which are allowable and the minimum state of serviceability required for the operation of the aircraft.   Each individual operator might prescribe more stringent requirements prior to the clearance of the aircraft.

Items essential for safety must therefore be kept well-maintained. The standard of maintenance on non-essential items, however, may well be much lower, dependent upon airline policy.   Galley equipment and service carts are often to be found in a state of neglect, and such a situation can lead to accidents (M27).   Service cart wheels which fail to caster freely, for example, can cause the cart to topple;   faulty brakes leading to unexpected movement of the cart can result in burns, scalds and bruising.   Loose, or ill-fitting items in the galley lead to similar accidents particularly in the presence of turbulence.   A dislodged oven, for example, being both hot and heavy can cause serious damage.

Whilst the cost of maintenance will be a conspicuous item in the company accounts, costs arising from poor standards of maintenance will normally be concealed.   For this reason alone, operators should take care to avoid "economies" in maintenance, which could be counter-productive.

ATMOSPHERE

Passengers and cabin crew share the same cabin atmosphere. However, in contrast to passengers, cabin crew spend their working lives exposed to this atmosphere and are engaged upon physical activity requiring higher respiration rates for most of the period of exposure.

Pressurization

An altitude of about 10,000ft is the maximum which can be safely and comfortably sustained without respiratory support. In the early years following the Second World War, the experiments and trials of the 1920s and 1930s came to fruition in the form of pressurized cabins in regular airline service. This development allowed the piston-engined aircraft of the time to fly at altitudes above much of the weather. With the advent of the turbo-jet engine, the fuel consumption of which precludes economic flight at lower levels, pressurization of transport aircraft became universal.

The Environmental Control Unit (ECU), generally referred to as a "pack", provides the facilities governing pressurization, ventilation, and temperature control in the cabin and the flight-deck. Outside air, having passed through compressors in the jet engine, is fed through flow-control valves at high pressure and a temperature of about 200°C. After being cooled, and at reduced pressure, the supply is piped to various sections of the aircraft interior.

Automatic control systems regulate the amount of high pressure air to be admitted. With aircraft altitudes up to about 20,000ft, the cabin altitude can be maintained approximately at sea level. As the aircraft climbs above 20,000ft, the cabin altitude will fall at the rate of about 500ft/min such that it reaches a value of about 8,000ft when the aircraft is at its maximum altitude. During descent, a reverse process occurs with the rate of change in cabin pressure being reduced to about 300ft/min in the interest of comfort. The flight-deck crew have the facility to adjust the rate of change within certain limits. The effects of failure in the pressurization system are discussed under "Decompression" in Chapter Seven.

Relative humidity

Since the perspiration and respiration of personnel on board provide the principal sources of water into the atmosphere, an important characteristic of the passenger cabin is low relative humidity. A comfortable level of relative humidity is around 40%-45% in winter and around 40%-60% in summer. It is necessary for the packs to remove water from the air at low altitudes in order to avoid condensation. As the aircraft climbs, the air becomes dry with the result that the relative humidity in the cabin, which is a function of ventilation rate, passenger load factor, temperature, and pressure, will typically vary between 2% and 23% (N03).

When relative humidity is below 30%, the mucous membranes become dry and this can result in considerable discomfort in the eyes, nose and throat. This may lead to painful irritation, difficulties in speech and in swallowing, and to lowered resistance to infections such as colds. A survey of work injuries and illnesses reported by flight attendants during 1979 showed that flight attendants had twenty times the expected frequency of respiratory illness compared with other workers (D05).

Reduced relative humidity also has a deleterious effect on skin, accelerating the ageing effects.

In an atmosphere where there is a low level of humidity, individuals must consume sufficient amounts of liquid to avoid dehydration. Extreme dehydration may lead to kidney and liver diseases. Passengers who consume large quantities of alcohol will have an increased tendency to dehydration as alcohol has a diuretic effect, causing the elimination of fluids. More relevant to crew is that coffee is also a diuretic. Because of their high level of physical activity, flight attendants are more vulnerable than passengers to the effects of low relative humidity. A fluid intake of about 0.5 l/hr would normally suffice.

Ventilation
A supply of sufficient oxygen is essential to replace that which is used by crew and passengers in the process of breathing. This requirement can be satisfied with very low levels of flow. In order, however, to achieve an atmosphere which is pleasant and acceptable, higher levels will be necessary. With rates of air movement below 0.05m/s (10ft/min), a feeling of stagnation is created. Such a rate may be insufficient to remove odours and airborne contaminants, and it may be difficult to achieve adequate control of cabin temperature. A stable vertical gradient will be established such that the temperature increases with height above the cabin floor. People favour a reverse gradient with the feet being slightly warmer than the head.

Penalties result, at the other end of the range, if the rate of air movement is in excess of 0.3m/s (60ft/min). Uncomfortable cooling draughts will be felt on the exposed parts of the body. At altitude, an increase in the amount of outside air taken into the cabin will increase the ozone levels and will also reduce the relative humidity. Rates of air movement must therefore be controlled between these upper and lower comfort levels (N03).

The ventilation rate is expressed in volume of air per passenger. This value will vary greatly with load factor. Within the average rate calculated for an aircraft there will also be variations in different sections if multi-class seating density is in operation. Typical values are in the range 0.003 - 0.02m^3/s (6-42ft^3/min). Higher rates are to be found on the flight-deck due to the cooling requirements of electrical equipment in an environment subjected to high levels of solar heating.

There are economic reasons which encourage airlines to keep ventilation rates as low as possible. One study indicated that a saving of about 1% on the fuel bill could be achieved by reducing the ventilation rate in a McDonnell Douglas DC-10 from 0.008 - 0.004 m^3/s (N03). As fuel accounts for something in the region of half the operating cost, such a saving is of considerable significance. Since the packs use bleed air from the engines, both performance and maximum payload are reduced by the use of the ECU. Further cost penalties result from the power used to cool the air, the drag induced by the heat-exchange units and, of course, the weight of the equipment.

Technical developments in the cooling processes employed have led to a need to mix the incoming outside air with warm cabin air before using the mixture in the cabin. Filters remove some, but not all, of the contaminants, but the oxygen depletion remains in the re-circulated air. This energy-saving use of re-circulation seems likely to escalate in the future.

Ventilation and cigarette smoke
Environmental tobacco smoke (ETS) comprises a complex mixture of many gases and small particles, the concentration of which increases with the amount of smoking and decreases with the ventilation rate of outside air.

Numerous difficulties stand in the way of dealing effectively with the ETS issue. First, the assessment of the concentrations of the various constituents presents technical problems. Secondly, there remains a great deal of uncertainty about the relation between concentrations, long-term exposure, and the extent, if any, of possible damage to health. Thirdly, complaints about smoking from passengers and cabin attendants may bear little relationship to the real effects of ETS.

Several studies have been made of concentrations of ETS constituents in aircraft cabin air (eg M02). These have shown that the concentrations of carbon monoxide and of nicotine are well below those prescribed in any official standards. Until such time as a good deal more is known about the effects upon health of exposure to the constituents of ETS it remains impossible to establish credible standards which could be applied to the aircraft cabin (C25).

One reason for the low measured concentrations of atmospheric pollutants in the cabin is the high ventilation rate which is necessary to produce acceptable air quality for the comfort and well-being of the occupants. The relative humidity, however, is normally much lower than that recommended for comfort, and will decrease as the duration of the flight increases. This can lead to irritation of the eyes, and to dryness and soreness in the throat. Closely similar symptoms are produced by exposure to ozone. Such symptoms may well be wrongly attributed by passengers and cabin attendants to ETS.

Industrial experience of Human Factors has on many occasions shown that complaints about an aspect of the work situation, such as noise or temperature, become more frequent when the work-force is disgruntled about a totally unrelated matter. Since many passengers are known to suffer high anxiety levels concerning flying, it may well be that ETS serves as a focus for complaint.

Prohibitions upon smoking in the cabin have been applied in the US, Australia, Canada, and elsewhere. Airlines benefit in several ways including the cost savings deriving from reduced ventilation rates and the elimination of minor damage to seats and furnishings caused by cigarette burns. The careless disposal of cigarette ends, particularly into waste containers, has been cited as a possible source of several in-flight fires.

More research is necessary to clarify the relationships between ETS and health before the aviation industry, together with all other facets of society, can adopt a properly well-founded position in relation to the effects of what has been called "passive smoking".

Ozone

Ozone, a form of oxygen having a different molecular structure from that of the more commonly occurring constituent of the air, is more prevalent at the higher altitudes favoured in the interest of fuel economy. Jet transport aircraft typically cruise at levels between 25,000 and 40,000ft, whilst supersonic transport aircraft fly above this range. At levels above the tropopause, the surface which defines the boundary between the troposphere and the stratosphere, ozone may be present in quantities which are harmful. The height of the tropopause varies both with latitude and with the seasons of the year between about 26,000 and 55,000ft. Supersonic aircraft will typically cruise above the tropopause as will subsonic aircraft during, for example, a polar flight in winter. Thus ozone, introduced into the cabin from the outside air through the Environmental Control Units, is a potential hazard for a large number of passengers and crew. It is fortunate that due to the high temperatures involved during compression in supersonic flight, most of the ozone is converted to normal oxygen before entering the cabin of Concorde. Ozone levels can be reduced by passing the air through catalytic absorption devices.

The effects of exposure to ozone include difficulties in breathing and inflammation of the lungs. In addition, sore throat, nose bleeds, chest pain and fatigue may be experienced. Symptoms typical of ozone-toxicity were reported three-to-four times more frequently by flight attendants in aircraft flying long distances at high altitudes than by those in aircraft flying short distances at lower altitudes (R10). Flight attendants are more exposed than

passengers to the effects of ozone because they are more active
and therefore have a higher respiratory rate.   In addition, they
breathe less humid air than seated passengers and this may increase
the effects of ozone (P08).

The major effect of ozone is to restrict air flow in the
bronchioles and this is associated with coughing.   Symptoms appear
at levels of 0.2 - 0.3 parts per million by volume (ppmv) of ozone in
the air.   Occasional peak levels of 0.57ppmv in the cabin of the
Boeing 747 have beem observed on the polar route, with average
values of 0.02.   With the benefit of humidification,   no symptoms

Table 3.1   The effects of ozone

Parts per million by volume (ppmv)	Effects
0.01	can be smelled
0.1	no apparent harm
0.2 - 0.3	easily smelled
0.25	[FAA limit in aircraft cabins]
0.5	distress
5.0	stupefaction and possible death

were experienced by passengers or crew (P08).

The Federal Aviation Administration (FAA) limits the ozone concentration levels to an average of 0.1ppmv with peaks of less than
0.25ppmv.   This is considered "innocuous to healthy humans for
indefinite exposure", though it is questionable how this might affect
individuals who are not healthy, particularly those suffering
pulmonary disease (M18).   Indeed, in the light of more recent press
reports, it appears that this level may be too high for healthy
individuals and that lung damage may be caused by ozone
concentrations lower than this level (G13).

Cosmic radiation
While cosmic radiation has an effect at heights above ground level,
it is not considered a serious health risk at flight levels typical of
subsonic transport (N03). However, the risk increases with increasing
altitude.

The supersonic passenger transport aircraft, Concorde, is fitted
with a warning system to alert the flight crew to the presence of
undesirable levels of cosmic radiation and accordingly to prepare to
descend to a safer level (P08).

Because exposure to radiation has a cumulative effect, it poses a greater threat to flight attendants and to flight crew than to passengers, and should be considered as a factor in relation to total exposure to radiation, such as medical and dental x-rays, and should be avoided by pregnant women. The maximum permissible dose (MPD) determined by the International Commission on Radiological Protection was 0.5 rem/year for the general public and 5 rem/year for radiation workers. For passengers, this level would be attained only after 61 supersonic round trips across the Atlantic in one year. The average annual dose for cabin crew in Concorde over a three-year period was 0.225rem (P08).

AMBIENCE

Lighting
A great deal of information has been gathered concerning the levels of illumination required in interior environments (I01). The required level depends upon the type of task to be performed, the reflecting qualities of the surroundings and the age of the people using the area. For tasks such as reading well-printed documents, a level of about 300lux would normally suffice for younger people. A higher level might be required for older passengers. In the aircraft cabin, this can be provided by way of general roof lighting supplemented by individually-controlled reading lights. At times when the cabin is darkened to enable passengers to sleep, a level of about 50lux is required for safe movement about the aisles.

Difficulties created from differing requirements of adjacent passengers are only partly accommodated by the use of individual lights. Similarly, cabin crew may need to have their workplaces illuminated causing spillage of light in the direction of passengers wishing to sleep. Analogous problems occur in relation to noise levels. Cabin crew members often need to employ a tactful approach to achieve acceptable degrees of compromise on these matters.

Noise
Sounds are produced by airborne vibrations impinging upon the eardrum. Unwanted sounds are the weeds of our acoustic environment, and are termed "noise".

The subjective loudness of a sound is a function not only of its acoustic power, but also its frequency. For this reason, the units most commonly used to measure noise levels, described as "A" scale, are weighted to compensate for the constituent frequencies within the sound. The decibel scale used is based upon the logarithm of the acoustic power, since very large changes in the power values are necessary in order to bring about perceptible changes. Since the pressure of sound waves is more easily measured than the power, the units employed are derived from pressure measurements and called Sound Pressure Level (SPL), the common unit being the decibel (dB). When the A scale is employed, the units are expressed as dBA. Typical SPL values of various sounds are illustrated in Figure 3.3.

Noise has three deleterious effects. First, it can cause damage to hearing. Even a brief single exposure to levels in excess of 130dBA, as might occur in the vicinity of a jet aircraft at high thrust levels, can result in permanent hearing loss. Longer-term exposure to lower levels can have the same effect. In the cabin of a typical propeller aircraft, about eight hours per day would constitute the limit of allowable exposure to avoid risk of damage.

Secondly, noise can interfere with communication. Adjacent passengers in a jet airliner cabin during cruise may need to raise their voices slightly in order to converse. At a distance of 2m (6ft) in a piston-engined aircraft two people would be obliged to shout loudly in order to be understood.

Thirdly, noise can cause annoyance and discomfort. There are large differences between people in their reactions, but the levels in modern transport aircraft are acceptable to a majority of

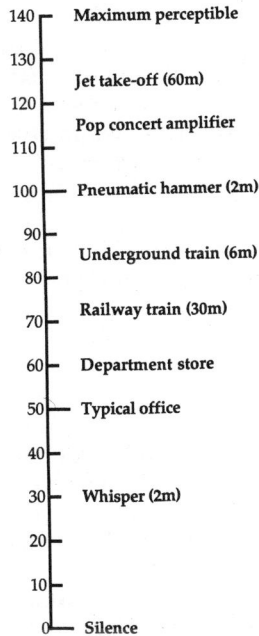

```
140 ┬ Maximum perceptible
130 ┤
    │   Jet take-off (60m)
120 ┤
    │   Pop concert amplifier
110 ┤
100 ┼── Pneumatic hammer (2m)
 90 ┤
    │   Underground train (6m)
 80 ┤
    │   Railway train (30m)
 70 ┤
 60 ┤   Department store
 50 ┼── Typical office
 40 ┤
 30 ┤   Whisper (2m)
 20 ┤
 10 ┤
  0 ┴── Silence
```

Figure 3.3   Sound levels expressed in dBA. Distances between origin and reception are shown for point sources.

passengers. It should be noted that a certain background noise level
in the region of 60dBA is required to provide the degree of privacy
that most people require in order to converse comfortably.

In light aircraft, and to some extent in large piston-engined
aircraft, engine exhaust accounts for a great proportion of the
noise. Propellers and fans generate aerodynamic noise emanating
principally from the tip. In a jet aircraft during cruise the primary
source of noise is the turbulent boundary layer of air produced by
the disparity between the speed of the aircraft and the relatively
stationary air surrounding it.

These noises outside the aircraft are transmitted through the non-
rigid skin and enter the cabin. There they must be absorbed by
wall-panels, seats, carpets, and the passengers themselves if accept-
ble levels are to be achieved.

Noise levels vary in different sections of the cabin due in part
to the location of the engines and also to the increasing depth of
the boundary layer from the front to the rear of the aircraft.
Rear-engined aircraft are therefore generally less noisy than those
with wing-mounted engines although high levels of noise may occur
in the rear of the cabin. In jet aircraft, the front section of the
cabin is normally the quietest place.

Vibration

In the past, vibration was a problem for cabin occupants. Since the
advent of jet aircraft, this is no longer the case. Flight attendants,
however, are liable to suffer more from the effects of the vibration
in the floor of the cabin because they spend so much of their time
walking about the aircraft in the course of their duties. The
vibration in the floor is transmitted to their feet and legs and this
results in increasing permeability of the capillaries. Fluid thus
leaks out of the capillaries and causes swelling and discomfort in
the feet and legs (G05).

SOFTWARE

As the passengers enter the aircraft cabin, they will be aware that
the airline has provided a range of Hardware facilities for their use,
and has staffed the cabin with personnel trained to tend to their
needs. The Software components of the system, that is the rules,
operating procedures, and practices, will be less conspicuous and can
easily be overlooked. Without this Software, chaos would rule in
the cabin, which would soon become both uncomfortable and
dangerous.

In order to keep the cabin serviced, cleaned, and provisioned for
the flight, plans will have been put into operation to bring together
all the necessary resources. Crew scheduling procedures will have

ensured that the correct number of staff of the correct type has been allocated to the flight. Company regulations ensure that the cabin attendants are properly prepared and equipped to fulfil their assignments.

The aircraft captain is, of course, in command. Whilst captains retain overall responsibility for the conduct and safety of the flight, certain duties, such as informing passengers about the safety equipment, are delegated to the cabin crew. Under the pressure of contemporary operating practices, it is unusual for the captain to be able to spend much time upon the supervision of activities in the cabin, with the result that the responsibilities falling upon the senior cabin attendant are considerable. Nevertheless, it is wise practice for captains to make themselves known to all crew members and to demonstrate interest and involvement in all aspects of the operation (O02).

The senior cabin attendant (sometimes called the purser, chief steward, or even the flight director) will brief the crew concerning any special features of the trip. A quick check of the crew's knowledge concerning safety equipment and procedure may be carried out. Cabin duties, including the locations to be manned in an emergency, will be allocated to each of the attendants taking into account seniority levels.

The purser's briefing will, of necessity, be concise. In order to allocate specific cabin duties, for example, it may be necessary to do no more than to display names alongside numbers on a list. The significance of such numbers will be defined within the airline's Software for the particular aircraft type.

Throughout the flight, most of the crew's tasks will be carried out according to well-known and frequently-rehearsed routines. It would be impossible to attend to the needs of passengers, and to complete the various duties, in any other way. It is of the essence of Software as a system resource that decisions are made well in advance of their implementation and can therefore receive careful and detailed attention in an atmosphere free from time constraints and other sources of stress.

Standard procedures are taught during basic training of personnel. Most companies produce a cabin crew training manual, of which each crew member will retain a copy. This document contains the knowledge requirements of the job. A separate handbook will be carried by cabin attendants on duty, in which essential information is set out, much of it in tabular or diagrammatic form. There will probably be a plan diagram of the cabin indicating the location of all emergency equipment and the duty stations of crew members. Lists will indicate items to be checked prior to embarkation of passengers, and further lists will show the duties to be performed at other stages. There will be notes about the special needs of disabled passengers and advice about first aid and elementary

medical assistance.   A comprehensive set of lists will define the
procedures to be followed in the event of an emergency.   This
documentation serves as a custodian of the system Software.

It has often been thought, quite erroneously, that Human Factors
(HF) is concerned only with the Liveware-Hardware relationship,
often described as "man-machine interaction". As outlined in
Chapter One, and further developed in Chapter Six, HF addresses
more generally the role of Liveware within systems.   Two broad
groups of problems arise in relation to the Liveware-Software
interaction. The first is concerned with the rules and procedures
themselves; the second arises from the necessity for these to be
recorded and communicated.

Operating procedures should be formulated such that they match
the capabilities and limitations of human performance. Consideration
should be paid to the "natural" response in a given set of
circumstances and the procedures developed, as far as possible, from
that.   Rules should be formulated by reference to a shared
knowledge base such that any particular rule may be understood as
the most appropriate choice from a set of options.   Observance of
such guidelines will lead to a high level of conformity, but in the
event of violation, steps should be taken to determine whether the
Software is in need of revision.

The benefits of optimal Software design can readily be negated by
poor standards of promulgation. Associated documentation must
therefore achieve high standards as a communication medium in
addition to satisfying the criteria of precision and completeness.
The correct representations of information must be selected to suit
both the user and the task.   Practical examples include clarity of
wording, page layout, and type face; the selection of appropriate
diagrammatic and tabular formats; rational sequencing and indexing
of material; convenience in availability and use (W10).

REVIEW

The aircraft cabin should be designed and operated in such a way
that safety and effectiveness are achieved.   Being a densely
populated part of the aircraft, the cabin makes heavy demands upon
the technology of Human Factors. The sizes and shapes of people
are highly relevant, as are their atmospheric requirements together
with considerations of lighting, noise and vibration. The seat, a
passenger's most intimate point of contact with the aircraft,
requires attention. The needs of the cabin staff must equally be
met in order to permit efficient task performance and to minimize
the occurrence of traumatic injury and long-term damage to well-
being.

The person-centred approach of HF must be integrated within a broader framework of system design in order for viable design solutions to emerge. The SHEL model provides one way of conceptualizing the interactions between system resources as they come together to form a practical and functional whole.

4 The Passengers

A typical passenger load will comprise people of different shapes and sizes, the young and the aged, the fit and the infirm. Native languages and cultural backgrounds may differ. A few people may exhibit undesirable behavioural characteristics. An understanding of such variety will assist cabin staff to cope with the problems which arise.

THE VARIETY OF PASSENGERS

It was estimated in 1984 that 70% of all Americans over the age of 18 had flown at least once in a commercial transport aircraft (G03). Thus, while it is probable that aircraft passengers will exhibit the range of characteristics typical of the general population, these characteristics may well not occur in the same proportions. For example, just under half of all trips are for the purposes of business; passengers travelling on business are predominantly male, since women comprise only about a quarter of this group. Business travellers and the small proportion of passengers drawn from the lowest socio-economic groups serve as examples of bias.

Passengers will vary in age from infants to elderly people, and in physical dimensions from the normal range to extremes of height and weight. Some may be disabled; some may be seriously ill. Passengers may travel alone or in groups of various sizes. The passengers and flight attendants may share the same native language or there may be more than one mother-tongue on board.

Some of these travellers may have special needs. Unaccompanied children and elderly people are of particular concern to flight attendants. Some passengers may require individual attention because of physical disabilities; some may have problems arising from fears and anxieties associated with flying. Others may, for a variety of reasons, behave in ways which are disturbing to other passengers.

AGE AND PHYSIQUE

Children

In common with coach and car interiors, the aircraft cabin is not ideally suited to the needs of infants and young children. For young children able to move about independently, restrictions upon their movements may cause discomfort to themselves and annoyance to those in their vicinity. However, allowing children to run about the cabin may cause problems of a different kind; they may be a danger to themselves and to others. Long journeys may exacerbate problems of restlessness in young children and boredom in older ones. Infants who travel at reduced cost because they do not occupy a seat must remain on an adult's lap throughout a flight if no spare seat is available.

Unaccompanied children between the ages of six and twelve years may travel in the care of the cabin crew; unaccompanied children below the age of six years are not normally accepted as passengers by airlines.

There are problems in relation to emergency situations where small children and babies are concerned. The seat belt fitted to the aircraft seat will not restrain a young child whose centre of gravity is different from that of an adult. Special restraints have been designed for children up to 1.02m (40in) in height but these may not always be available. Larger children and small babies remain a problem for which no special solution has been devised.

In the event of a crash landing, there are particular difficulties in safeguarding babies. The infant may be held on the adult's lap, supported by the adult and protected to some degree by the adult's body. Whilst it is possible for an adult in a "brace" position to hold a baby, the force of the impact may cause the baby, even when belted, to be thrown from the adult's lap. Alternatively, the infant may be placed, well wrapped, on the floor behind a bulkhead. However, this is only feasible in those emergencies in which there is sufficient advance warning. Neither solution is completely satisfactory (J13).

There are specially designed life jackets for infants weighing less than 35lbs (16kg), and for children weighing up to 90lbs (41kg), but these are not always available. Babies can be accommodated in buoyant cots.

Following a decompression, there is a tendency for responsible adults to attend first to the well-being of the children in their charge, relegating their own safeguarding to a secondary consideration. This is a grave error in relation to the donning of oxygen masks where the adult may become disabled if attention is not immediately paid to obtaining oxygen, and thereby be incapable of aiding the child. The same principle applies, albeit with less force, to life jackets.

Elderly passengers

Elderly passengers are likely to suffer from impaired vision and hearing and thus may benefit from individual pre-flight briefings. Due to their frailty, some elderly passengers may need special attention in an emergency. Their lack of muscular strength and restricted movement may lead to difficulties in operating the seat belt and in moving at an adequate speed to the exit in an emergency. These passengers may have difficulties in accessing and donning life jackets. The reduced atmospheric pressure in the cabin may lead to mild hypoxia in elderly people and a consequent reduction in the level of cognitive performance and an increase in confusion. Elderly individuals tend to have longer reaction times and thus may respond more slowly to signals of the presence of danger: they may therefore require help in order to don oxygen masks expeditiously. Some elderly people may have difficulties in remembering instructions because of deficits in short-term memory and may become confused and easily distracted in an emergency, requiring continual direction in order to make progress to the exit.

Demographic trends

There is evidence to suggest that the most quickly-growing segment of the airline market is the age group 55 to 85 years, which includes those with sufficient time, income, and motivation to travel. This group is sometimes referred to by market researchers as Woopies (well-off older people). The implications of this trend for passenger management during normal flight and during emergency situations are considerable, bearing in mind that compared with younger groups, this group is likely to move more slowly, to be more forgetful, and to have more difficulties in seeing and hearing.

Physique

The design of the cabin interior may be better suited to the size and shape of some individuals than of others. Individuals who deviate markedly from the normal range of stature or of girth are likely to suffer discomfort in flight. Regulations require that, for design purposes, seat occupants are assumed to weigh 170lbs (77kg), a weight which will be exceeded by many obese passengers and larger males. This weight parameter also applies to the design of seat belts. Seat belt buckles may be uncomfortable for obese or for pregnant passengers. Extension belts have been designed to reduce these problems. Constraints on leg room are encountered by those of above average leg-length and because seat pitch increases with the grade of passenger accommodation, discomfort is likely to be greatest in low-cost travel.

Restricted seat pitch also has implications for safety. The adoption of the conventional brace position may be difficult if not impossible for some passengers, such as those who are tall and those who are obese. Emergency egress will, for all, be hampered by close seat pitch, whilst obese passengers may be particularly disadvantaged in an emergency evacuation through being wedged in their seats.

HANDICAPPED PASSENGERS

The central problem regarding the carriage of handicapped passengers is the need for these passengers to be assisted during an emergency evacuation of the aircraft. This has in the past discouraged or even prohibited handicapped passengers from utilising commercial air transport. During the 1970s, however, there was a move to reduce discrimination against handicapped people in a wide range of activities so that they might enjoy the same facilities as able-bodied individuals.

With respect to air transport, the FAA commissioned a study of the emergency evacuation of handicapped passengers (B13). This showed that it was useful to distinguish between ambulatory and non-ambulatory handicapped individuals, the latter usually requiring a wheelchair for mobility. Successful evacuation depends on the ability to move through the cabin at the rate of at least 0.3m/s (1ft/s). While 96% of the ambulatory handicapped individuals in the study were shown to be able to reach an exit 29ft 2in (9m) from a window seat within 30s, only 37% of those who were non-ambulatory could reach the exit unaided within that time, and would thus require assistance in the event of an emergency evacuation. This led to the FAA definition of a handicapped person as "one who may need the assistance of another person to expeditiously move to an exit of an aircraft in the event of an emergency".

The study demonstrated that handicapped individuals occupying window seats spent up to 50% of the time to reach the exit in moving from the seat to the aisle. Total evacuation times for groups were shorter when non-ambulatory handicapped individuals were seated away from the exits. The study underlined the contribution of narrow aisles and restrictive seat pitch to the difficulties of assisting non-ambulatory handicapped individuals. It also showed that assistance in operating the seat belt was necessary for most of those participating in the study.

Data of this kind are in themselves neutral and may be used by regulatory authorities to support the continuation of a prohibition of the carriage of non-ambulatory handicapped passengers or, alternatively, for indicating the sort of problems to be overcome by design philosophies and operational procedures if handicapped passengers are to be adequately accommodated. The tide of public opinion was strongly in favour of such accommodation and the FAA ruled that the requirement for the assistance of another person in order to evacuate in the event of an emergency was not sufficient grounds for refusing to carry a passenger.

The American Civil Aeronautics Board in 1982 ruled that all passengers regardless of handicap should be given reasonable access to commercial air transportation and the opportunity to use ordinary unaltered services of airlines. "Only significant and clearly demonstrable safety concerns or the most extreme considerations of carrier inconvenience should justify the refusal to carry handicapped passengers." The International Air Transport Association (IATA) procedures state that handicapped ("incapacitated") individuals will be carried provided that they do not jeopardise the safety of other passengers, that they comply with specific conditions applicable to their transportation and that the airline is provided with sufficient medical information about the passenger's condition.

As a consequence of these rulings, passengers with a variety of physical disabilities may be encountered in flight. These disabilities may be of a permanent nature such as blindness, deafness, paraplegia, crippling diseases (for example, muscular dystrophy and arthritis), or they may be such temporary conditions as broken limbs. Pregnancy, while not a disability in the clinical sense, may be a handicap when travelling by air. Pressure changes may have a deleterious effect and pregnant women are not usually accepted as passengers on international flights after the 35th week of the pregnancy, nor on domestic flights after the 36th week.

Handicapped passengers must be briefed individually before the flight and the briefing must be tailored to the passenger's disability.

Ambulatory handicapped passengers

This group includes those who are obviously handicapped such as blind people and obese people; and also those less obviously handicapped such as deaf people.

Blind passengers are frequently able to move about independently provided that they have the assistance of their guide-dogs or canes, both of which may be brought into the cabin and located close to their owners.   However, canes are potentially hazardous in an emergency and must therefore be securely stowed prior to take-off. Flexible canes must be collapsed and may be stowed flush with the cabin floor under the window seats, or in enclosed overhead compartments.   Rigid canes may be stowed under seats in the same row as long as they do not protrude into the aisle, or stowed between the window seat and the side-wall, but not in an emergency exit row.   The use of canes and crutches in a simulated emergency evacuation, however, did not improve escape times. These aids could prove hazardous for other passengers during movement through the cabin and could, in addition, cause damage to the escape slide (B13).

If a blind individual is accompanied by a guide-dog, then it is advisable for them to remain together. In an emergency evacuation down a slide, the blind person and guide-dog should be evacuated together so that no time is lost once on the ground in moving quickly away to safety.

Safety briefing cards will not be helpful to blind people and some appropriate method is necessary to ensure that the information on the cards is conveyed to the blind passengers. Solutions include the use of Braille briefing cards (Air France, for example, provides these for blind passengers on longer flights), "hands on" experience of emergency equipment (oxygen masks, life jackets), and counting the rows to the emergency exits. It is also necessary to instruct blind passengers on the method of fastening and releasing the seat belt.

The problem for deaf passengers is the requirement for information to be received visually. If they are seated at the back of the aircraft, they will be better able to see what is happening in the cabin and be in a good position to follow visual instructions from the cabin staff. However, problems may arise during emergency situations when environmental conditions reduce the effectiveness of vision. Following a decompression, there is a sudden fogging of the atmosphere; fire gives rise to smoke; the failure of all electrical systems on impact could result in sudden darkness. In all these circumstances, the ability of the deaf person to receive information is severely reduced.

Non-ambulatory passengers

Those passengers who are unable to board the aircraft or to move about without assistance are likely to be wheelchair users. More than 40,000 chair-bound passengers are, for example, carried annually by British Airways (A03).

Because of their batteries, powered wheelchairs are classified as Dangerous Goods which may be excepted from the provisions of the Dangerous Goods Regulations when carried by a passenger. This exception is conditional upon the operator's approval and with the proviso that non-spillable batteries are disconnected and the battery terminals insulated to prevent accidental short circuits. Wheelchairs with spillable batteries must always be carried and loaded in an upright position and the battery, which must be firmly attached to the chair, disconnected. The captain must be informed of its location in the aircraft. If the chair cannot be maintained throughout in an upright position, the battery must be removed and the wheelchair is then carried as checked baggage without restriction.

These conditions have caused concern to wheelchair users who consider that their wheelchairs are not well cared for and too often sustain damage during transit (N28).

The American Civil Aeronautics Board regulations require air carriers to make "reasonable efforts" to allow folding wheelchairs to be taken on board and stowed in the passenger compartment, if this does not violate other regulations. If the wheelchair is not carried in the cabin, then it is desirable that it should be made available promptly at the door of the aircraft, so that passengers may use their own wheelchairs as much as possible.

Travel for chair-bound passengers has improved somewhat since the 1960s when forklift trucks were commonly used to transport disabled passengers to the cabin. Loading bridges and apron transports now permit the wheelchair to be taken as far as the entrance to the cabin.

However, there remain problems for these passengers. Typically, passengers' wheelchairs are too wide to negotiate the aisles of an aircraft and while an in-flight wheelchair may be available to permit mobility during flight these chairs have not been considered satisfactory by those with experience of their use. An impressionistic survey showed that in-flight wheelchairs were considered inadequate on account of their small wheels, having no wheels (thus requiring the passenger to be carried in the chair), having no arms, and being of a small size (N28). This suggests that a redesign of the in-flight wheelchair, taking into account all the requirements of chair-bound passengers while they are engaged in the total flying context including the use of the chair in the terminal, would be desirable.

Non-ambulatory passengers have particular needs in respect of access to and use of the lavatories which are not always adequately met. There are considerable space requirements if the wheelchair is to be used effectively to provide access to the toilet compartment. The design of the interior of transport aircraft may be decided to some extent by the airline purchasing the aircraft and thus there is no standardization in the dimensions of aisle width or in the location of the toilet compartments. There are frequently restrictions on the space available for manoeuvring wheelchairs because of the location of partitions, of flight attendants' seats, and of narrow aisles and doorways. Limitations on space may require the disabled passenger to prepare for the use of the lavatory while still outside the compartment, and to leave the door open during the transfer from the chair to the lavatory. Solutions involving curtains or the use of interconnecting mid-cabin toilet compartments have been proposed.

Individual pre-flight oral briefing of non-ambulatory passengers (including their attendants) must specify the routes to each appropriate exit and the most appropriate time to begin moving to an exit in the event of an emergency. Information must be sought from the passenger regarding the method of assisting him or her so as to avoid causing pain or further injury.

There are problems in assisting a non-ambulatory handicapped person in an emergency evacuation. There is very little space in which to manoeuvre because of fixed arm-rests, restrictive seat pitch, and restricted aisle width (B13). Passengers must be orientated correctly for the slide in order that they do not exit head first. Repositioning the handicapped passenger for the slide increases the total time taken for evacuation.

Some airlines adopt the practice of placing a blanket in the seat to be occupied by a non-ambulatory handicapped person to facilitate emergency evacuation. However, this practice was not welcomed by those disabled passengers surveyed concerning their flight experiences as it could be interpreted as a method of avoiding the effects of incontinence (N28).

Seat location for handicapped passengers

There are two considerations to be taken into account in relation to the seat location of handicapped passengers. The first is the safety of all those on board and the second is the comfort of the individual handicapped passenger. Unfortunately, these two considerations may conflict. The seats most likely to accommodate a stiff or plaster-encased leg are those near the doors and emergency exits, and these are not available for passengers whose mobility is in question, as they might impede an emergency evacuation.

Passengers who are hemiplegic should be seated with their non-paralysed side adjacent to the aisle to facilitate their movements out of the seat in an emergency evacuation.

A constraint on passengers requiring therapeutic oxygen is that they sit at least 3m (30ft) from the smoking area of the cabin.

Experimental evidence suggested that passengers with handicaps of any kind should be advised to sit in aisle seats (B13). Those sitting in the window seat took much longer to reach the exit than did those sitting in the aisle seat. However, there is the problem of able-bodied passengers in-board of the disabled person for whom the disabled person may be an obstruction to their reaching the aisle.

The study also suggested that even ambulatory handicapped passengers should not be seated in a row adjacent to a door or overwing exit where they may impede the evacuation or be injured in the rush of other passengers. These seats are more suitable for able-bodied passengers who would, in an emergency, be able to open the door, to assist other passengers, and to look outside to judge whether it was safe to use the exit.

The same study indicated that those passengers requiring assistance in order to evacuate should be seated away from congested areas where crowding by other passengers may interfere with those who are preparing them for evacuation. Tests in which incapacitated passengers were represented by anthropological dummies showed that when these were located far from the exit, 33 passengers evacuated the cabin in the first 45s whereas when they were located near the exit, only nine passengers had evacuated within this period.

Other handicaps
Some intending passengers are intellectually handicapped. This condition may be a consequence of old age, injury, or disease; it may be congenital. People handicapped in this way are often unable to look after themselves, particularly in an emergency. They are easily distracted from the task in hand and are unlikely to follow instructions adequately.

Some intending passengers are psychotic. They may be quiescent or they may be sufficiently disturbed that their behaviour inconveniences and upsets other people, and their condition renders them incapable of looking after themselves.

An airline will not accept as a passenger any person who will cause inconvenience or discomfort to other passengers, or jeopardize the safety of the flight, or jeopardize his or her own health. These conditions might not be considered to apply when the person is accompanied by another passenger who is responsible for the disabled passenger. In cases of doubt, an airline may require medical clearance before accepting the individual as a passenger.

MEDICAL PROBLEMS

Illnesses encountered among aircraft passengers may range from slight malaise to those which culminate in death. It was estimated in 1980 that for every million passengers, a life-threatening medical emergency to one of them would necessitate an unscheduled landing for treatment (M22). Given an air-travelling population of 318 million during the previous year, this represented about six incidents per week. With the growth of air travel in the intervening period, the incidence of unscheduled landings might be expected to have increased. Birth, as well as death, has taken place in flight.

Flight-induced medical problems
Perhaps the most obvious of the minor illnesses associated with flying is airsickness which is the manifestation of motion sickness in an aircraft. Those who are susceptible to other types of motion sickness, for example carsickness, are also likely to be susceptible to airsickness. Airsickness is partly due to the incompatibility of the sensory input from the eyes and from the organs of balance. While the latter are transmitting information that the individual is being shaken about, the information from the eyes, focused on a stationary interior, suggests that the individual is not moving. Anxiety is a further contributory factor. Once commonplace,

airsickness has diminished as cabins have become more comfortable and aircraft fly higher thereby avoiding most of the problems associated with rough air. There are age and sex differences in susceptibility to motion sickness (R09). Women are more prone than men to suffer from it; the peak age appears to be around ten to twelve years old; those who are middle-aged or elderly are least susceptible.

The cramped conditions associated with low-cost travel can place passengers at risk from deep vein thrombosis and pulmonary embolus. Conditions are exacerbated by dehydration due to the low humidity, and also by mild hypoxia which encourages inertia. Cases have been reported of the symptoms appearing during the flight and also of their delay for several days (C26). It was estimated that, over a three-year period, this was the cause of 18% of the 61 sudden deaths of long-distance passengers at London's Heathrow Airport, which is approximately three deaths per year (S07). The risks are best reduced by exercising the legs as much as possible during a flight and drinking non-alcoholic liquids.

Some common medical emergencies may result from incidents which take place during flight. These include, for example, choking, food poisoning, hypoxia due to decompression, injury as a consequence of turbulence. Some medical emergencies result from substance abuse where overdoses can lead to coma or even to death.

Pre-existing medical conditions
Some medical problems may arise from the exacerbation of a pre-existing condition, possibly due to mild hypoxia as a result of exposure to the ambient cabin pressure. This could lead to a heart attack in one who suffers from cardiac dysfunction. Other chronic states which may become acute are asthma and epilepsy.

Arthritic passengers may have particular problems in fastening and unfastening seat belt buckles. This could give rise to problems in an emergency evacuation when the belt has to be released quickly, or during turbulence when the belt must be fastened quickly.

Reactions to pressure changes
Some passengers will react adversely to changes in pressure associated with changes in altitude. These changes can cause discomfort and, in some cases, pain. This is due to the effects of changes in pressure on the gases inside the body. Passengers with defective dental fillings have been known to experience toothache. Increases and decreases in pressure outside the body are normally compensated by equalisation of the pressure within. However, those individuals with a history of middle-ear infections have difficulty in maintaining equal pressure between the middle ear and the ambient atmosphere, and are particularly vulnerable to earache

during descent as the cabin pressure increases. Those suffering from colds or other respiratory tract infections are likely to endure considerable discomfort, particularly during the acute phases of these infections, due to inequalities of pressure in the nasal sinuses.

Following the well-known physical laws, intestinal gases expand by 50% at a pressure corresponding to an altitude of 10,000ft and 25% at a pressure corresponding to maximum cabin altitude. While this is not a problem for normal healthy people, individuals with colostomies may require a larger bag in flight. Those with a spastic gastro-intestinal tract may suffer severe pain as a consequence of the distension of the bowel by the expansion of intestinal gases.

The bodily responses of passengers who smoke cigarettes and drink alcohol are altered as altitude increases. The effectiveness of the cells of the body to take up oxygen, which is essential for life, is reduced by alcohol. As the amount of oxygen available at higher altitudes is less than that available at sea-level, it follows that the heavy drinker is likely to be more vulnerable to the effects of altitude than the passenger who does not drink excessively. Such effects would make the person particularly vulnerable in the event of a decompression.

Those who are heavy smokers may also be vulnerable to the effects of altitude. Carbon monoxide, a constituent of tobacco smoke, combines with the haemoglobin in the blood with the result that the amount of oxygen in the body is depleted. A heavy smoker at sea-level may be affected in the same way that a non-smoker is affected at 12,000ft above sea-level (M13).

BEHAVIOURAL VARIATIONS

A survey of emotional behaviour in passengers based on reports of stewardesses from three major American airlines indicated that passengers behaved in the air in much the same way as they behaved on the ground except that, as a group, they tended to behave in a very dependent fashion (B25). This dependence may account for the reluctance of passengers to don oxygen masks or to take any other positive step in an emergency in the absence of explicit instructions. Young men in groups were considered more likely to create a disturbance than similar individuals travelling alone. The majority of examples of extreme emotional behaviour were concerned with fear (67 out of 156) followed by anger (57). One example of fearful behaviour involved a middle-aged male passenger who opened the emergency window and jumped out just before take-off. Instances of extremely angry behaviour were associated with complaints of poor service and excessive intake of alcohol.

The emotions of fear and anger have major implications both for the behaviour of the passengers experiencing them and for other passengers and flight attendants who are involved in the consequences.

Fear of flying

Some passengers may suffer from emotional distress such as claustrophobia (fear of enclosed spaces) or fear of flying, which in some individuals may give rise to a full-blown panic attack. In one instance, a pilot turned back to the airport shortly after take-off so that a passenger with excessive fear of flying could be deplaned.

It has been estimated that 8% of the population suffer from at least a mild phobia of some type (D02). Many of these people will display several phobias. A survey showed that 30% of Americans had never flown and it is likely that this percentage includes those whose fear of closed-in spaces or of flying is so extreme as to prohibit them from considering the possibility of air travel (G03). Fear of flying has been the subject of some concern among the airlines as it has implications for lost revenue. A study carried out in the United States in 1979 attempted to gauge the extent of fear of flying both in the general population and among those who had experience of flying (D03). The results indicated that more than two-thirds of the general population reported "no anxiety" in connection with flying, about 13% reported "anxiety" and about 18% claimed to feel "fear". The distinction between fear and anxiety was not made clear: presumably anxiety refers to a less severe level of negative feeling. Among those who had flown, more than three-quarters said they experienced no anxiety and about 10% said they were afraid. Expressed fear of flying was reported more by women than by men, and by the lower socio-economic groups compared with the higher ones. From these results it would appear that at least 25% of passengers approach their flights with some degree of apprehension. The results of a 1984 Gallup poll are shown in Table 4.1 (N26). It is possible that a reluctance to admit fear, particularly on the part of male participants, leads to an underestimate of its prevalence.

Table 4.1 Fear of flying (N26)

Level of fear	%
No fear	65
Sometimes	21
Most of the time	3
All of the time	11

Great care must be exercised in the design, administration, scoring, and interpretation of surveys of this type. Words such as "fear" and "anxiety" are ill-defined and can be used in many ways. Emotional issues are particularly likely to produce responses designed to conform with notions about "proper" attitudes and reactions.

Not all anxiety can be classified as phobic. Some can be based upon perfectly rational analyses of risk. Thus, for example, a person living quietly in rural surroundings may well be increasing, by a small amount, the risk of injury or death having decided to travel by air.

Those with extremely phobic responses to flying are unlikely to entrust themselves to aircraft. Under such pressures as job requirements or family bereavements, others decide to fly in spite of high levels of anxiety. These passengers are often only able to do so after implementing strategies to reduce their fears. Such strategies, in common with other phobic defence mechanisms, are essentially concerned with the avoidance of any stimulus or event which might increase anxiety and tension. Unfortunately they are likely to prove disadvantageous in the event of an emergency.

One such method of coping with fear of flying is to attempt to numb emotions with the use of alcohol and tranquillisers. This may reduce unpleasant feelings but only at the cost of depressing central nervous system activity. The level of arousal is reduced. In an emergency, when it is important to be alert and competent, the behaviour associated with alcohol or drug ingestion is likely to be disadvantageous for survival.

Another method is to pay selective attention to events in the environment, avoiding any communications which are concerned with the possibility of danger. Oral safety briefings by cabin staff at the beginning of the flight and the information on briefing cards carried in the cabin inevitably draw attention to the possibility of danger. Although tension may be reduced by the deliberate ignoring of briefings, such a policy leaves the individual bereft of the information required to react to an emergency.

If these strategies for coping with fear of flying are effective, the individual is not optimally fit for coping with a real emergency's demands of alert and competent behaviour. If these strategies are ineffective, then the individual's level of existing fear will be increased in the face of a real threat and this too is likely to disrupt adaptive survival behaviour.

Where fear of flying precipitates a panic attack, which may occur in 2% - 5% of the general population, the intervention of an appropriately trained flight attendant can provide some alleviation from distress, though in extreme cases, it may be necessary to return to the airport to deplane the passenger.

While there is an argument in favour of some mild anxiety in a strange situation, extreme anxiety is disadvantageous.

There have been numerous attempts to make available programmes designed to assist people to overcome their fear and anxiety. These have been presented either in book-form, for self-administered treatment, or as classes for groups of people to attend, usually organised by, or in conjunction with, airlines (Y01).

These programmes adopt several parallel approaches to the problem, loosely based upon well-established psychological principles of learning. First, education about flight is used in order to diminish the anxiety which arises from ignorance about aircraft and their operation.   Explanations are offered concerning the changes in engine sounds which can be heard, the rumblings of the gear being lowered or retracted, the use of wing flaps and spoilers, and other such items perceivable by passengers. The purpose is to remove the sources of unrecognised and fear-provoking stimuli and to attempt to train the individual to regard the cabin as a place with which he feels familiar.

Techniques such as forms of yoga, autogenic training, or progressive relaxation are used to assist people to reduce anxiety levels by means of physical and mental relaxation. Advice may be offered concerning ways of distracting the attention away from sources of anxiety and towards pleasant and desirable matters.

The group classes may then include progressive approach techniques which involve a step-wise movement towards eventual flight. Such schedules include visits to an aircraft cabin to encourage familiarity with the surroundings and culminate with a short flight. Every possible attempt is made to utilise the mutual support between members of the group. This camaraderie plays a large part in the dissipation of anxiety.

Cultural variations
Difficulties may arise as a consequence of cultural differences between groups of passengers, and between passengers and flight attendants. The snapping of fingers, for example, to gain attention may be acceptable in some cultures but offensive in others. The differing status of women in various communities may similarly create problems.

Whilst aircraft are used principally by people from the developed nations, others from less-developed countries will have little or no prior experience of some features to be found in an aircraft. This lack of technical sophistication may give rise to inappropriate or even dangerous behaviour.

International airlines whose routes encompass countries with differing culture patterns often include in their cabin staff training programme an appreciation of these differences and their implications for behaviour.

Deviant behaviour

It is a regrettable fact that passengers steal from aircraft. Consumables are stolen from the toilet compartment and, more seriously, items of safety equipment, such as life jackets and megaphones, are stolen. In fact, anything that is not securely attached to the aircraft is vulnerable to theft. This practice has been cited in objection to the provision of smoke hoods. More important than the cost of replacement is the absence of safety equipment in the event of an emergency.

Other forms of deviant behaviour may result from fear or distress. The most frequent sources of angry and unruly behaviour, however, are the excessive consumption of alcohol and the insistence on smoking at inappropriate times and in inappropriate places.

Passengers who are drunk have been abusive to other passengers and to flight attendants, sometimes to the extent of inflicting physical violence. It is unfortunate in this context that atmospheric pressure compounds the effects of alcohol such that at a cabin pressure corresponding to 8,000ft, one ounce of alcohol has the effect of two taken at sea-level. Excessive drinking of alcohol can cause passengers to vomit in their seats or in the aisle, with unpleasant consequences for other passengers and for flight attendants who must deal with these problems.

Infringement of the smoking rules is a second major source of friction. Passengers who smoke in no-smoking areas of the aircraft will upset other passengers who are likely to express their annoyance clearly to the flight attendants. It may be difficult for flight attendants to maintain order between the two parties and the situation could become very tense. Passengers may persist in smoking while walking about the aircraft, or in the toilet compartment. This is not only a violation of the regulations but it jeopardizes the safety of the flight. Some passengers will not accept any limitations on their smoking activities and their response to the flight attendants' intervention is verbal, and sometimes physical, abuse.

Less common but equally problematic is the unruly behaviour which is the consequence of substance abuse. Stimulants, depressants, and deliriant drugs all in their different ways have an effect on emotions and behaviour. Restless, agitated, violent and even psychotic behaviour may result from overdoses of stimulants, while overdoses of depressive drugs may lead to loss of consciousness. Overwhelming feelings of panic may accompany ingestion of some deliriant drugs. In extreme cases of overdose and withdrawal, an unscheduled landing may be required.

It may be necessary in the last resort for the flight attendant to call the captain to restore order to a situation that is out of control. The captain has the authority to restrain unruly passengers in handcuffs and to call ahead for police to meet the flight, or even, in the event of a serious disturbance, to make an unscheduled landing to deplane the offending passengers.

To the extent that it interferes with flight attendants' carrying out their cabin duties, unruliness in passengers constitutes a violation of regulations. Further interference with the duties of a flight officer results if a flight officer leaves the flight-deck to help resolve a disturbance in the cabin.

Pressing charges against unruly individuals, however, is expensive and may be considered by airlines to generate unwelcome publicity. For aggrieved passengers who may have been assaulted, there are difficulties in securing corroborative evidence from other passengers. The general problem of obtaining witnesses willing to testify in court is exacerbated on flights where potential witnesses may be travelling much farther onwards.

5 The Cabin Crew

Proper selection procedures and training schedules are necessary in order to ensure that cabin attendants are prepared for the demanding tasks required of them. Some resilience is necessary to cope with the unavoidable sources of stress which accompany the job. Long-term career prospects add a dimension which, in former times, was absent.

THE EARLY YEARS

When aircraft first started to carry passengers in the years after World War One, provision for the comfort of passengers was made on the ground before the flight. This included the distribution of such items as hot water bottles, cotton wool to put into ears, and leather coats. It was not long, however, before in-flight services were introduced. In 1922, British Daimler Airway was the first airline to employ a steward to serve refreshments on routes to continental capitals. Later, in 1926, Imperial Airways inaugurated the "Silver Wing" service between London and Paris during which a steward served drinks and lunch. With the introduction of flying-boats, a high level of sophistication of in-flight services was achieved where drinks and meals were cooked and served, and sleeping accommodation was available. These developments required the employment of stewards whose task it was to attend to the needs of passengers and provide services similar to those provided by a steward on a passenger-carrying ship. There were no safety connotations.

In the United States at this time it was the responsibility of the co-pilot to look after the passengers, ensuring that provisions were on board, greeting the passengers, passing around the boxed lunches and serving coffee from Thermos flasks (M33). It was the flight mechanic in the short-lived Philadelphia Rapid Transport Airline who attended to sick passengers and cleaned the interior of the aircraft

(D01).

The 1930s saw the introduction of air stewardesses, or "hostesses" as they were frequently known. The first to be employed were nurses hired by Boeing Air Transport in 1930 on domestic flights. While their function on board was to look after the passengers and tend to such problems as airsickness or minor accidents, they also had a public relations role in the campaign to popularize air travel against the competition of the railroad. In Europe, Air France first employed stewardesses in 1931 followed by KLM in 1934.

An early American flight attendants' manual advises the reader "to maintain the respectful reserve of a well-trained servant. Captains and pilots will be treated with strict formality while in uniform and a military salute will be rendered the Captain and pilot as they go aboard and deplane". The duties of these flying servants included keeping the window sills dusted and using a small broom on the floor after every flight. They were expected to "swat flies in the cabin after take-off" and to warn passengers against throwing lighted cigarette butts out of the window, particularly over populated areas. The instructions to carry a railroad timetable "just in case the plane is grounded somewhere" and "to walk passengers going to the washroom to see that they open the right door rather than the adjacent emergency exit" convey an authentic period flavour (K05).

There was initial reluctance on the part of the international carriers to employ women on long-distance flights but in 1943 Pan American Airways recruited its first stewardesses as a contribution to the war effort. A description of the ideal stewardess of this period as "blue-eyed with brown hair, poised and self-possessed, slender, 5 feet 3 inches tall, weighs 115 pounds, is 23 years old, actively engaged in some participant sport, an expert swimmer, a high-school graduate, with business training, and attractive" suggests a public relations role (B06). This impression is reinforced by the manner in which the British Overseas Airways Corporation (BOAC) was finally persuaded to employ stewardesses on the North Atlantic route in 1946, which in effect was compliance with pressure from the sales office in New York. These stewardesses were to be "British, preferably lightweight" and "willing to undertake the duties of their task which should be described to them as similar to those of a domestic servant" and within the age band 23 to 30 years. The stewardesses (this title was emphasized to eschew the "glamour and frivolity" thought to be associated with "air hostess") were required to undergo a ten-week training course at the airline's Catering Training School with particular emphasis on the "duties required of them as waitresses and elementary first aid" (B06).

Although recruited for public relations purposes, the stewardesses were faced with an arduous and demanding job. The hours were long. The working period from London via Gander and New York to

Montreal extended to 25 hours. Half of BOAC's flights to New York were more than 24 hours late in 1947 and duties of stewardesses included lowering bunks and making beds after serving six-course meals (B06). Nevertheless, there were satisfactions to be obtained from the job. Aircraft were small and it was possible to attend to individual passengers' needs. Because air travel was relatively expensive, airline passengers tended to come from the topmost socio-economic groups and on transatlantic journeys, a luxury atmosphere was provided in order to compete with the ocean-going passenger liners. On some inter-continental journeys, the aircraft, its crew and passengers all broke their journeys for overnight stops at exotic places. An unintended, but valuable, benefit of an overnight stop was that it provided for readjustment of the internal clock to changes in time zones and for rest and relaxation between stages of the journey.

SOME STATUTORY PROVISIONS

With the passage of time, the rules governing all aspects of air travel have increased and the job of the flight attendant has also become subject to regulations. The British Air Navigation Order states that cabin attendants shall be carried for the purpose of performing duties in the interest of the safety of passengers (Article 17). In the United States, Federal Aviation Regulations emphasize the safety function of flight attendants in emergencies and in emergency evacuations (FAR 121.391, FAR 121.397).

Regulations also specify the minimum number of flight attendants carried in an aircraft, which varies according to aircraft capacity. FAR 121.391, for example, states that at least one flight attendant should be carried where there are more than nine but fewer than 50 passengers, and thereafter one attendant for every unit, or part unit, of 50 passengers.

The Joint Requirements for Emergency and Safety Airborne Equipment of the European Civil Aviation Conference (Doc.18) state that for all types of aeroplanes having more than 19 seats, the number of required cabin attendants is one for each unit (or part of a unit) of 50 passengers on board: and that the minimum number of cabin attendants shall not be less than half the total number of Types A, I or II floor level exits.

They also state that "the number of aeroplane types in which cabin attendants are qualified at any particular time should be limited" (2.4b).

The location in the cabin of flight attendants during take-off and landing is also subject to regulations. These require that flight attendants be located as near as practicable to floor level exits and uniformly distributed throughout the aeroplane [eg FAR 121.391 (d)].

THE DUAL ROLE

There are two aspects to the job of the flight attendant. These are service and safety.   Training is often conducted by wholly different groups of instructors, sometimes at different locations.   The safety aspect is the subject of regulations.   In addition to the regulations cited above, there are regulations governing the content of the safety training programmes for flight attendants.   Regulations also prescribe certain activities to be carried out on every flight, such as the oral briefings at specific times during the flight.   The major function of the flight attendant, as legally defined, is to safeguard the passengers by providing leadership in times of emergency and by competently managing any potential hazards.   The appropriate behaviour in these circumstances is authoritative and commanding.   However, this function is rarely exercized.   Most of the time there is no serious turbulence; no major decompression occurs; fires on board are unusual; ditching and emergency landings are very rare events.   Thus flight attendants are infrequently called upon to use the skills which comprise the statutory necessity for their presence on board.

The service role which accounts for the greater proportion of the job performance involves caring for the passengers in a nurturing capacity.   The appropriate behaviour is compliant.   "Friendly, caring personalities" are specified in an advertisement for cabin crew.

The service role has been stressed by airlines in their advertising. The public relations function of stewardesses in particular is unchanged since their introduction.   Compelling evidence of this can be seen in the results of a Top Ten annual poll carried out by the magazine "Business Traveller" in 1986.   The Asian airlines were voted into the Top Ten and among the main reasons given by respondents were "the charming, perfectly groomed air hostesses who turn the tedium of a long-haul flight into the pleasures of a magic carpet" (M21).   These are appropriate images for personnel whose role is confined to service and to making the customer happy.

There is a certain tension between the two aspects of the flight attendant's role which may have implications for effective performance.   It may be difficult in an emergency for the passengers to change their perceptions of the flight attendant from a "charming hostess" or a "caring personality" to that of a figure of authority whose directives have weight and whose commands are to be promptly obeyed. Similar difficulties may be experienced by the flight attendant in switching the mode of interaction unless attention has been paid to this problem during training.

A relatively minor example of this conflict may be illustrated in relation to the management of unruly passengers.   One of the hazards faced by flight attendants in recent years is the hostile behaviour of some passengers, the expression of which may range from verbal threat and attempts at intimidation to physical assault, with or without weapons. The victims of the attack may be other passengers or the flight attendants. In either case, the flight attendant must respond firmly in order to curb the behaviour, but at the same time minimising the deleterious effects upon the good will of all concerned.

Stewardesses may also be the subject of sexual harassment. This anti-social activity may be the consequence of alcohol intoxication, of fear and stress, or, at its simplest, from sheer boorishiness encouraged in some cases by membership of a group.

It is considered likely that assaults on flight attendants are under-reported and that the problem is therefore larger than is generally known. The requirement to maintain good order may conflict with the philosophy of the customer always being right. Apart from being a serious problem in itself, it has implications for the outcome of emergency situations because the effectiveness of flight attendants in the management of safety procedures is threatened when they are seen to be vulnerable.

SELECTION OF FLIGHT ATTENDANTS

A survey in 1985 showed that in the United States, 85% of flight attendants were female, 61% were married, and 43% had children. Their median age was 34 years; 29% were aged over 40 and 22% were aged less then 30 years. The majority (80%) flew between 70 and 80 hours per month; 12% flew more than 85 hours per month. Nearly 40% had more than four years of college education; 16% completed their education at high school (N03).

The criteria which are typically specified to attract applicants as flight attendants are certain physical limits on height and weight; an age range usually between 20 and 30 years; a certain minimal level of education which may include knowledge of foreign languages; an ability to do simple arithmetic; a pleasing appearance; and qualities variously described as "caring", "warmth", "friendliness". Often experience in a job which involves relating with the public is regarded as advantageous.

These criteria may be more or less stringent depending upon the airline and upon the prevailing employment situation. Some airlines specify the ability to swim a certain distance as a necessary accomplishment; some bar those who wear spectacles, though contact lenses are acceptable. For some a higher than minimal standard of education is required, and greater emphasis is placed on physical attractiveness in some airlines than in others. A characteristic which is valued is the ability to mix easily with others and to work well in a team.

There remains sufficient glamour associated with the job of flight attendant for most companies to receive numbers of applications far in excess of vacancies.   Some applicants seek long-term career appointments, others are attracted by the prospects of relatively high income and the opportunity to travel widely during a short period of employment.   The industry finds that suitable candidates in each of these two categories can fulfil its requirements. However, it is likely that there will be quite severe "pruning" at this early stage of those applicants judged to be unsuitable for engagement.

There is no uniform procedure in general use for assessing the suitability of candidates whose applications have resulted in their being selected for the next phase of the process.   Typically, an invitation is issued for candidates to spend a day engaged upon a variety of individual and group test procedures.   The former probably includes a medical examination, written tests of basic educational achievement, and possibly some psychological assessment of personality type.   The group tests are usually designed to evaluate the way in which an applicant reacts and cooperates with colleagues in a team, and provides an opportunity for several experienced people to assess the candidates' social skills.   Such programmes frequently end with individual interviews during which further detailed information may be elicited, and the interviewers make further assessments of the candidates' appearance and self-presentation.   Following these programmes, joint decisions are usually made during a conference of the assessment team.   Airlines differ in the relative emphasis placed upon the various factors being assessed, reflecting the particular needs and the current philosophy of the employer.

Language
English is the language of aviation.   Pilots and air traffic controllers usually communicate through the medium of English and an ability to understand and speak English is a necessary qualification for flight attendants on international routes.   For those whose mother-tongue is not English, therefore, fluency in both English and their own language is essential.   Those whose mother tongue is English may or may not be proficient in a second or third language.   The national level of foreign language expertise is, of course, reflected in the cabin staff.   Compared with the inhabitants of continental European countries who learn languages other than their own from an early age, native English speakers are relatively rarely fluent in another tongue.   Advertisements for cabin staff by a British airline state that "ideally" a candidate will have conversational ability in one or more foreign languages, whereas continental carriers may require four or five.

Airlines are required to brief passengers in safety procedures and

emergency evacuation. To comply with this in the case of non-
English-speaking passengers, most United States overseas carriers
require their flight attendants to be bilingual in English and the
language of the country to which they fly most often.
Alternatively, ground personnel from that country come on board
before take-off and carry out the briefing in their native language,
or a taped briefing is played with accompanying gestures. The
briefing card is intended to require no language in order to be
understood.

In the event of an emergency, however, there are likely to be
problems when the flight attendants cannot communicate directly
with the passengers.   A McDonnell Douglas DC-10 was chartered
to carry 364 Moslems who did not speak English. There was one
interpreter on board and the flight attendants did not speak Arabic.
The aircraft crashed in Turkey and a post-crash fire started. There
were considerable difficulties in communicating with the passengers,
most of whom were unaware of the fire, that carry-on baggage
which included holy water and sacred rocks should be left behind
and that shoes must be taken off (B07).

In the light of the evidence that a majority of passengers do not
attend to the pre-flight briefing and less than 10% read the briefing
card, the value of direct and unambiguous instructions at the time
of an emergency is apparent (F14).

TRAINING

Training periods vary between about four weeks and four months
depending upon the airline. The two main topics to which training
is devoted are safety and service. In addition, the training may
involve some ancillary topics such as familiarization with the airline
industry, the design of the aircraft flown by the company, the time
zone changes, and weather. Stress is laid on the importance of
personal grooming. Airlines with major international business may
also include within their training schemes some appreciation of
cultural differences.

Training for service
The competent flight attendant needs to have knowledge ranging
over a wide spectrum of topics. Passengers are liable to expect
answers to their questions concerning the airline and its routes,
immigration and customs procedures, some local details about the
destination, connecting flights and surface transportation, and all
other matters relating to a particular flight. Whilst relevant
literature is carried on board, the flight attendant may be required
to draw upon a broad base of general knowledge.

Similarly, a wide range of skills is required to ensure the comfort of passengers throughout the flight.   Airlines are aware of the commercial value of a favourable image and of the extent to which cabin crews contribute to that image.   Passengers are welcomed aboard, helped to find their seats and stow their luggage.   Even such apparently simple tasks can induce a measure of stress when account is taken of the fact that some passengers may be disabled, others may be angry or afraid, and others disgruntled about the allocation of seats.

On short flights with a high load factor, the amount of time available to complete the various passenger services may be very small.   On the other hand, passengers travelling First Class on a long-haul flight will expect bar and meal service to rival that offered by a high-class restaurant.

Tax-free goods are available on board and can normally be purchased using any of several forms of currency or credit cards. Sales are highly profitable to the airlines and cabin staff may be paid commission.   Before the end of the flight, all the associated bookkeeping must be completed.

Service training is normally carried out using a mixture of class room and practical sessions.   In view of the high-revenue usage of aircraft, it is unusual for these to be available for training. Extensive use is made of simulation to allow trainees to learn and practise the required skills using mock-ups of cabins in which high levels of fidelity may be incorporated.   It is usual for fellow students and training staff to play the role of passengers during training exercises.

Throughout training, emphasis is placed upon personal appearance and presentation since such matters will be influential in the effectiveness of the service provided.

Training for safety

While training for service is at the discretion of the individual airline, and may take weeks or months depending on the range of topics addressed, the minimum acceptable standard of training for safety and emergency procedures is prescribed by the regulatory authorities and may take about seven days of the initial training course.   The objective of this training is to ensure that flight attendants are able to use all the emergency and survival equipment and to carry out all the emergency procedures and drills, including aircraft emergency evacuation (FAR 121.417 and CAP 360 13-16). Cabin crew should also receive instruction on normal flying duties "including the location and use of all galley and cabin equipment". In Europe, training is also provided for survival at sea, in uninhabited terrain, and in extreme conditions such as the Arctic or the desert.

Initial training covers topics such as First Aid, crew coordination, fire and smoke training, water survival training, handling drunk passengers and the allocation of seats in an emergency.

In addition to the general curriculum, training on particular types of aircraft must also be carried out both for newly employed staff and for those converting to a different type of aircraft. This includes such topics as the location and use of all emergency and survival equipment carried on the aircraft, extinguishing a fire and the use of protective equipment for dealing with smoke, and training for emergency evacuation.

There are regular statutory checks on the ability of cabin crew to perform their safety-related activities. In the UK, for example, recurrent training must be provided in preparation for the emergency survival tests which take place at intervals of 13 months. The CAA requires that, once in three years, cabin crew must practise the use of escape slides; the operation of each type of exit and emergency exit in manual, normal and emergency mode; extinguishing a fire; and the use of protection equipment against smoke where appropriate. The FAA requires that emergency drills are rehearsed every 24 months and that, every third year, cabin fire and smoke comprise the major topic addressed in annual checks.

Training methods vary considerably in different airlines. For example, training for evacuation from a smoke-filled cabin may range on the one hand from asking the question "what would you do in a post-crash fire?" to requiring flight attendants to manoeuvre through the aisles of a simulated cabin filled with "smoke" to reach the emergency exits and locate missing passengers (C22).

The procedures to be followed by the cabin staff are included in a manual, a copy of which is given to every flight attendant and is available on every flight.

Simulators

Due to the prohibitive cost of keeping aircraft available for the training of cabin crews, simulators are in extensive use for this purpose. The principle of simulation is based upon a degree of similarity between the skills which may be acquired and rehearsed in the training device, and those required in the performance of the task being trained

A cabin simulator may comprise a representation of the complete cabin, or one specific part of it such as a galley or a door. Individual items used within the simulator, such as fire extinguishers, oxygen bottles, ovens, or coffee machines may be identical with those to be found in the aircraft.

It is common practice for trainees to acquire their knowledge and skills by behaving sometimes as "passengers" and sometimes as "crew". Under the direction of an instructor, the various facets of cabin service and of emergency procedures are rehearsed in a realistic, but economic, fashion. Facilities for the simulation of

Figure 5.1 A double-width escape slide used for the training of flight attendants

emergencies include variable cabin lighting, sound effects, and smoke. Evacuation rehearsals will include the opening of emergency exits, deployment and use of escape slides, and the marshalling of personnel briefed to act the role of passengers. Some simulators are fitted with such refinements as sound level sensors to ensure that trainees' voices achieve the appropriate loudness when issuing verbal instructions.

In recent years, two modifications to more traditional training practices have emerged, both of which owe their origin to developments in the training of flight-deck crews. One of these, an augmentation of the simulator hardware, is motion, which provides the kinaesthetic sensations of the motion experienced in flight from taxi-ing, through take-off and cruise to descent, landing, and even crash landing. The second, a procedural shift in emphasis, involves the simulation of a complete flight or large sections of a flight, rather than of isolated emergency incidents. This is usually described as "Line Oriented Flight Training" (LOFT) and has demonstrated the benefits to be derived from training crews to manage incidents as a team, making the most effective use of available resources.

Using these two techniques together gives rise to highly effective cabin crew training schedules. Trainees act in turn as passengers in the simulator, some of whom will have been briefed to behave in a certain manner when the emergency arises. Following completion of activities on the ground, take-off is simulated with realistic sound and motion cues. The flight attendants go about their routine tasks. Suddenly an "emergency" develops. Commands "from the flight-deck" are given by recorded messages. There may be failures in lighting or smoke in the cabin, and other auditory, visual and kinaesthetic cues are introduced. When the simulator eventually comes to rest, it may well be in an unusual attitude. Emergency evacuation is initiated by taped commands from the flight-deck and the trainees acting as flight attendants must then manage the evacuation which may be interrupted by the failure of a slide at an exit door.

Contrary to popular belief, realism in simulation is not of itself a desirable objective. The important criterion in the evaluation of any training device is the extent to which transfer of training takes place, that is, that skills learned during training are effectively performed in the real situation. Evidence is necessary to determine the characteristics of training devices and procedures which best facilitate transfer. It can be shown that some facets of simulator fidelity fail a test of cost-effectiveness. Experience has shown, however, that behavioural fidelity is likely to be essential for transfer to occur, that is to say, trainees should not rehearse behaviour patterns which differ from those required in the skill being taught. An example might be taken from the disposal of the hatch from an overwing exit. If, during simulation, there is some reluctance to throw this heavy object outwards and clear of the escape route, similar behaviour patterns, contrary to the procedure laid down, may be elicited under the stress of a genuine emergency.

PROGRESS

Assessment
Flight attendants are required to demonstrate their statutory knowledge and skills to the satisfaction of the regulatory authority, which normally delegates the task of assessment to the training

organization, subject to inspection of its facilities. Over and above this basic requirement, an airline is likely to set its own standards regarding the trainees' competence in non-statutory duties.

The assessment of knowledge is a relatively simple matter by the use of written tests and oral questioning. The standard attained in the performance of skilled tasks is more difficult to assess. Trainees may be required to demonstrate the use of galley equipment, or to serve a meal to "passengers". A judgement about the level of competence may then be made. Similarly, demonstrations may be required in the operation of safety equipment, and the management of personnel during a simulated emergency. In the most sophisticated simulators, realistic physical conditions may be reproduced and devices are available to monitor such variables as sound levels in order to ensure that flight attendants rehearse the use of appropriate voice volumes.

In spite of psychological selection tests and the use of efficient training techniques, it is difficult to predict how an individual will behave in a real emergency. Obviously, emergencies cannot be rehearsed and major ones will never be encountered by the majority of cabin crews. The role of motivation is important here to ensure that the cabin crews are not only able to perform in the appropriate way but that they are also willing to do so.

When training has been successfully completed, assessment must be made of on-the-job performance. This is carried out by supervisors. Sometimes, representatives of the company fly incognito to observe the way the flight is managed and to provide more information upon which an assessment can be based. More informally, assessments can be influenced by the letters received from passengers, either in praise or otherwise of a member of the cabin crew. While these letters may provide a valuable insight into the performance of a flight attendant, they should be used with caution in assessment and always discussed openly with the crew member concerned in order to provide an opportunity to correct any misleading statements and to comment on any negative remarks.

Career structure

The average length of service of cabin staff in the United States increased from around two years during the 1950s to about 11 years in the 1980s.

In some airlines, a one-year contract forms the initial stage of the flight attendant's career. This in effect allows for an extended assessment of the candidate's suitability for more permanent employment. It may also satisfy the needs of those employees who do not want to be committed to a longer period.

From the ranks of junior cabin staff, flight attendants can progress through an intermediate level, managing two or three others in a section of a large aircraft, to becoming the senior flight attendant on board, supervising as many as fifteen other staff. An added dimension of the job is the involvement of experienced flight attendants in both the selection of new staff and of their training.

Further progression leads to management positions involving selection policies, the design of training programmes, and the general administration of cabin affairs.

Some airlines have a policy of providing part-time employment to experienced flight attendants whose full-time career has been interrupted by family responsibilities.

Motivation

It is generally recognised that a highly motivated workforce makes a major contribution to the success of an enterprise.

Motivation operates initially at the time of recruitment. The attractiveness of the job will motivate potential employees to apply and, having been selected, to remain in the job. High levels of staff turnover are a drain on resources since recruitment and training are expensive operations.

Once in post, people must be motivated to carry out their tasks and to meet at least the basic standards required by the organization. Because no job can be specified precisely, there is normally some scope for an employee to respond creatively to the problems which might arise and to use initiative in the performance of tasks. Whether or not an employee will do so depends on the level of motivation. Job performance can therefore range along a continuum from compliance with basic standards at one extreme to exceptional efforts beyond reasonable expectation at the other.

The duties of a flight attendant consist largely of service to passengers but within a context of potential hazard; they are carried out away from home base and are not, for the most part, closely supervised. Given these circumstances, how can flight attendants be motivated to perform at levels far above that of minimal compliance?

Motivation is influenced by extrinsic factors and intrinsic factors.

Extrinsic factors are external to the individual. They operate to modify behaviour by means of an external feedback loop and are often described as "the carrot and the stick". Extrinsic factors include the pay and conditions attached to the job and such benefits as travel concessions. Benefits equally available to all employees are effective mainly in attracting individuals into the job and ensuring a level of performance sufficient to remain in it. They do not usually act as motivators for a high level of performance.

Specific financial rewards, or other extrinsic rewards such as promotion, are often used to motivate individual employees. Where it is possible to measure performance and where employees understand the operation of the reward system and perceive it to be equitable, then extrinsic rewards can be effective motivational tools.

However, objective measures may be difficult to achieve.

It is also important that the provision of rewards does not reinforce one activity at the expense of others which may be equally significant but less easily measurable. On many flights, cabin staff are involved in selling goods and are often rewarded with a commission. If the sales activity is highly valued by the company, then it may be emphasized to the detriment of other aspects.   There may be less time available to attend to individual passengers' needs. Flight attendants may feel demoralized by such an implicit contradiction between a training that emphasises service and the reality of selling goods for a commission.   Such unintended consequences of a reward system must be considered in calculating its overall effectiveness.

Intrinsic factors operate by means of a feedback loop that is internal to the individual and are associated with such phrases as "self-actualization" and "pride in the job". Intrinsic motivation is derived from a sense of personal achievement in carrying out a job to a high standard.

The value to an employer of an intrinsically-motivated workforce is clear: minimal supervision and a willingness to work beyond the limits of the job specification. The value of intrinsic motivation to the employee is also clear: the satisfaction which results from exercising responsibility, using initiative to overcome difficulties, and performing the job well.

It is not a straightforward matter, however, to establish and maintain an intrinsically-motivated workforce. First of all, the people selected must be those seeking their vocational satisfactions primarily in the central tasks which make up the job rather than in the extrinsic rewards associated with it.

Secondly, the training of flight attendants must itself demonstrate the operation of a high level of intrinsic motivation on the part of the trainers. The philosophy of "do-as-I-say-but-not-as-I-do" will be totally counterproductive: trainers must expect that their example will serve as a model for trainees.

The objective of training is to internalize the high standards which have been taught. There are specific formal methods which can be used to inculcate the practice of critical self-evaluation. For example, after each exercise is completed, trainees should be encouraged to carry out a systematic review of each element of the tasks. This self-evaluation can then itself be reviewed with the trainer. When trainees begin flying duties, the practice of self-evaluation can be pursued by means of a formal de-briefing after the flight in which performance in the job is reviewed and evaluated. In this way, trainees are taught to think about the manner in which they do their jobs and how they can improve their performance. Training along these lines will encourage the development of an internal feedback loop which is sustained by habits of self-evaluation and self-criticism.

However, selecting the appropriate recruits and training them to be internally motivated does not alone guarantee high-level performance. There are also organizational factors to be considered. These include management style, the reward system, the extent to which employees can participate in making decisions which affect them directly, the channels of communication through the hierarchy, and the organizational climate. All these factors play a part in determining the extent to which internally-motivated behaviour will be facilitated.

COMMUNICATION

Effective communication is an important, albeit neglected, skill in most walks of life. For the flight attendant, the ability to communicate effectively contributes significantly to the service function provided to passengers, but is no less important either in normal crew team-work or during emergencies. Different styles are required to match the different circumstances.

Communication takes place when a message is transmitted from a sender to a receiver. Normally, written or spoken language serves as the medium to facilitate this transmission. Verbal messages are accompanied by gestures, vocal intonation, posture, and facial expression, all of which make up the non-verbal components of the message. It is impossible to eliminate non-verbal components from verbal messages. Even during an announcement via the PA system, non-verbal signals such as breathing pattern, tone of voice, or hesitation all contribute to the totality of the received message. Simple messages can be transmitted by means of non-verbal communication alone.

Non-verbal communication plays an important part in conditions where noise or the distance between those communicating is too great for verbal communication to be effective. This is particularly relevant in emergencies. Working groups often develop their own language of non-verbal communication which enables them to work more effectively together. It is also used in situations where expressing opinions or reactions verbally might infringe rules or conventions. There is anecdotal evidence of flight attendants whose patience has been excessively tried by a particularly demanding passenger "accidentally" tripping with the result that soup or coffee is spilled into the passenger's lap.

For communication to be effective, that is for a message to be correctly received and understood, it is important that the non-verbal components are properly attuned to the content of the verbal material. Where these are in conflict, the impact of the verbal message will be considerably reduced if not negated. For example, giggling or embarrassment displayed by a flight attendant during the pre-flight briefing will have the effect of debasing the safety message, and conveying the impression that the topics addressed by the briefing are unimportant and unworthy of attention.

The effectiveness of the communication is also influenced by the extent to which the person originating the message understands the characteristics of the receiver. Clearly, a message delivered in the Japanese language will be wasted upon a listener who understands only French. But the problem is far more subtle than this. A native speaker might employ vocabulary and figures of speech unfamiliar to a person with only a working knowledge of a language. An expert in a given subject area might employ "jargon" terms unknown to the non-expert. Similarly, non-verbal gestures adopt a meaning and significance which varies with culture. The competent communicator will be well aware of such difficulties and will adjust the language style, speed of delivery, and message format such that the effectiveness of communication is optimized.

Having received a message, a person will evaluate its contents by taking into account the status and credibility of the sender. These will be determined to some extent by previously held beliefs. A passenger, for example, will assume that messages from the aircraft captain concerning the progress of the flight are reliable and accurate. In less structured situations, the credibility of a speaker may be determined by his appearance, accent, and style of delivery. Conflicts can be generated; credibility will be lost if communication techniques are poor. A flight attendant offering information in an unduly hesitant or confused manner, or whose statements fail to match with obvious facts, will lose the confidence of passengers.

Communications involving cabin crew during a flight may conveniently be grouped into three types; namely those with the flight deck, with passengers, and with other cabin crew colleagues.

The senior flight attendant is normally the person who acts as a link between the flight-deck and the cabin crew. Reports will be required by the aircraft captain concerning the number of passengers who have boarded, the readiness of the cabin prior to taxi-ing, and similar matters. Further advice about the facilities required at the destination, such as wheelchairs for disabled passengers, will be passed to the flight-deck for onward radio transmission. All such messages are to be conveyed in a concise and accurate way. Operational advice such as that concerning likely delays, weather conditions, or the risk of turbulence will be passed from the flight-deck to the cabin crew and sometimes supplemented by an announcement from the flight-deck crew by means of the public address (PA) system. In the event of an emergency, the purser will be briefed when possible so that the necessary preparations might be made. These matters are described in Chapter Nine.

In view of the high workload on the flight-deck during certain phases of a flight, unnecessary and untimely interruptions by cabin crew should be avoided. Some airlines have introduced rigorous rules concerning the extent to which cabin crew are allowed access onto the flight-deck. The most vulnerable periods are those between the beginning of taxi-ing until the aircraft is well clear of the ground, and the corresponding phase during the latter parts of the descent, approach and landing. In view of the potential danger resulting from distractions during these periods, nothing should be allowed to disturb the flight-deck crew from carrying out their operational tasks. This principle of the "sterile cockpit" has been incorporated into the US FARs.

Some mutual understanding between the flight-deck and the cabin concerning the duties of each is necessary to ensure smooth working practice. Should a passenger become critically ill just as the aircraft is leaving the ground, for example, then it becomes necessary to communicate this information to the captain at a time when, under normal circumstances, interruptions would be forbidden. Inquiries for information from the flight-deck immediately following the initiation of a go-around (missed approach) procedure, on the other hand, demonstrate a lack of understanding of flight-deck tasks. Analogously, the purser would prefer not to be summoned to supply coffee to the flight-deck at peak moments of activity in service to passengers.

Training in communication skills during most cabin crew courses is directed primarily towards interaction with passengers. Problems arise not only from varieties of language and culture but also from the necessity to cope with passengers who are excessively anxious, angry, intoxicated, or fatigued.

The third form of communication is that between members of the cabin crew. Each crew comprises a temporary group of attendants assembled for a single duty period. The amount of prior acquaintance may be little or none. The level of communication necessary for individuals to coalesce into an effective team is facilitated by the shared background of training and experience. Messages can be brief and in coded form.

The leadership style of the purser is one of the most important determinants of crew interactions. Communication is part of a complex social structure in which each individual's motives, in spite of uniformity in training, determine the pattern of behaviour adopted. Inexperienced junior attendants might, for example, be reluctant to report anxieties concerning some aspect of their work if they believe that there is some risk of censure or ridicule. This could be injurious to effective team-work and prejudicial to safety. The solution is largely in the hands of the senior cabin attendant.

SITUATIONAL FACTORS

Isolation

In two ways, the job of flight attendant involves an element of isolation which is absent from most other types of employment. First, the attendant's only contact with the airline for months on end is via immediate colleagues on board the aircraft. This small group, under the leadership of the senior attendant, remains intact for perhaps a single day, but seldom for much longer than a week. Secondly, there is the detachment from the more usual social and domestic aspects of normal lifestyles which are compatible with a stationary place of employment.

Obviously, the extent of each category of isolation varies according to the type of operation upon which the person is engaged. The flight attendant working on short-haul day time domestic routes from a company's main base may be little different from ground-based company employees. Long-haul operations, however, introduce conditions which are unusual. A working day may well begin in a hotel room thousands of miles from home. The surrounding culture, climate, and language may be wholly alien. Fellow crew-members provide the only link with the employer, and the day might end in surroundings as remote as those in which it began. For some, such variety may provide interest and enjoyment. For others, when the novelty has waned, it may be a source of stress, particularly in view of the difficulties in maintaining stable social relationships consequent upon the periods away from home. Senior flight attendants require, in the absence of any form of supervision, a particular resilience to the effects of occupational isolation.

Working hours

In a guide to requirements for the avoidance of excessive fatigue in aircrews, the UK CAA stated that the average weekly total of duty hours for cabin staff should not exceed 55; that there should be 2 consecutive days off duty in any consecutive 14 days, and a minimum of 6 days off duty in any 4 consecutive weeks. Cabin crew should be rostered such that their minimum rest period would be not more than one hour shorter than that of flight crew and that the combined standby time and the following Flight Duty Period (FDP) should not exceed 21 hours (C12).

In the United States there are no statutory limits to the hours of work of flight attendants. In 1984, 80% of flight attendants flew between 70 and 85 hours per month (N03). Those limitations which are in operation have been determined by the process of collective bargaining. In 1986, the FAA was petitioned by flight attendant unions to establish maximum limits and minimum rest periods, or at least to apply the 1985 rules for pilots on domestic flights, which guarantee a minimum number of hours of rest between trips. The outcome of the petition is expected to decide whether any change in the present situation is necessary.

The pace of work
The service of meals, and more particularly the sales-related activities, have implications for the pace of work. For example, on one flight where this was measured, 50% of the time of the cabin staff was concerned with taking purchase orders, fulfilling them, and distributing the purchases to the passengers (M27). With high density seating on a flight of about two hours' duration, each flight attendant would have available only three minutes per passenger to carry out the full range of cabin services. In-flight service is an area where airlines compete with one another and the tendency to increase the range and number of services to be provided in flights of short duration may tax severely the ability of flight attendants to cope. Flight attendants may be seriously fatigued at the end of a working day and not in an optimal state to cope with any emergency which may arise.

Sleep loss and sleep disturbance
Short-haul night flights require that the crew work during hours when they would otherwise sleep. Attempts to make up for this sleep loss during the day often cause difficulties as they run counter to the body processes which are undergoing a waking phase. In addition, there are practical and social difficulties which arise when people try to reverse the normal lifestyles associated with day and night. Since the majority of such flights carry cargo rather than passengers, this problem is less pressing for cabin crews than for flight-deck crews.

Long-haul flights interfere not only with the normal work/sleep pattern but cause additional difficulties on account of the time zone changes which are the inevitable accompaniment of long-distance east-west travel.

Many bodily processes operate on a cycle which corresponds in duration to the period of the earth's daily rotation (H11). These include sleep, digestion and elimination. There are also less obvious changes associated with circadian rhythm and these include changes in body temperature, in hormonal levels and in the composition of the blood. Some time-oriented functions such as the monthly menstrual cycle operate on cycles longer or shorter than twenty-four hours. All these cycles are disrupted by transmeridian flight and take different periods to readjust. "Jet-lag" is the popular short-hand term to describe these effects of sleep disturbance and time zone changes on bodily and mental functioning. There are wide individual differences ranging from no apparent effect to almost total collapse for a day or more. In general, the transmeridian effect appears to be less severe when an individual flies from east to west than from west to east.

Whilst it is impossible to specify precisely the period required for resynchronization to occur, the rubric of one day per hour of time-shift serves as an approximate guide. Thus it is likely to take about five days to restore equilibrium after a five-hour time zone change such as that involved in a flight from London to New York. This presupposes that after a flight, the individual is able to remain in the same time zone gradually adjusting to the prevailing time-determined conditions. Such circumstances are not those of crew members whose working life involves constant change such that adjustment of cycles one to another and of the whole to a particular time zone may only rarely be achieved.

A major effect of sleep disturbance is fatigue which is manifested in the deterioration of mental functions such as attention, perception, motivation, and decision-taking. This leads to decrements in performance resulting in errors and accidents. Continued exposure to sleep disturbance may lead to stomach disorders, constipation, and variations in the menstrual cycle. In addition, it is thought that the effects of the stress involved in sleep irregularities may lead to premature ageing effects (S10).

While sleep disturbance has deleterious effects on health and on job performance, some attempts to combat these problems are arguably more damaging to the individual, both in the long and short term. These include taking amphetamines to stay awake and alcohol or tranquillising drugs to go to sleep.

A survey of nearly 300 flight attendants conducted in one international airline showed a substantial increase in the use of sleeping drugs when on duty (46%) over their use when off duty (13%). Alcohol was said to be used as an aid to sleep by 45% of the sample. Little awareness of the interaction of these drugs was apparent among those interviewed (H03). This survey highlighted both the magnitude of the problem of sleep disturbance and the nature of the attempts to solve it. It is a problem likely to persist as economic constraints will not permit scheduling which is compatible with body rhythms. Other proposed solutions, including the use of relaxation techniques, have yet to be fully evaluated in the aviation environment.

With aircraft such as the Boeing 747 being capable of non-stop flights in excess of 20 hours duration, arrangements become necessary for the provision of crew rest facilities on board. Some recently fitted interiors include bunks to allow crew members to sleep during some part of the flight. It must be borne in mind, however, that some people experience difficulty in sleeping on aircraft. Scheduling procedures should take proper account of the wealth of information available on human requirements for rest and sleep.

Injury and illness
In spite of technological advances in aviation, the cabin as a workplace for flight attendants has many disadvantages. The dry atmosphere has an ageing effect on the skin and increases vulnerability to respiratory infection. The working area is cramped as a consequence of the priorities afforded to space and weight. Shortcomings either in design or in maintenance expose flight attendants to the risk of injury.

In the event of an accident to the aircraft, flight attendants are more likely than passengers to be injured. Their "stations" at exits make them liable to injury on impact from the collapse of partitions or coat-closets; during turbulence flight attendants are four times more liable than passengers to suffer serious injury; being physically active, they are more at risk than are sedentary passengers from the effects of decompression.

In short- and medium-haul flights, the working period of flight attendants is characterized by almost ceaseless physical activity. This includes walking, stooping, bending, reaching, pulling, carrying, jerking and lifting. These activities must be carried out within a very tight time-scale, and for the most part in full view of the passengers. Thus there is a need to maintain a professional interface with the passengers while carrying out physically demanding tasks. In this context, the finding is not unexpected that nearly 70% of the disabling occupational injuries reported by a group of flight attendants comprised fractures, sprains and strains (D05). Back injuries, accounting for more than 25% of the total injuries, are the largest single group reported by flight attendants (G05).

There is some evidence suggesting that the incidence of illness in general is less for flight attendants than for other comparable workers. However, they showed an increased incidence of respiratory disease, diseases of the inner ear, and aerotitis, due possibly to the reduced humidity and higher ozone levels encountered in flight. Aerotitis, the failure of the middle ear to adjust for pressure changes through the Eustachian tube, was found to be more prevalent among female than male flight attendants (W08). In one airline, immunity to hepatitis B was found to be higher for male flight attendants than for pilots or female flight attendants (H09). Oedema in the legs may occur as a consequence of extended periods of standing (C01).

Uniform
Clothes intended to be worn for work should ideally be designed in relation to the tasks to be carried out while they are being worn. Some thought should also be given to the maintenance and care of clothing in relation to the standards of appearance required in the job.

There are two aspects of the flight attendant's uniform which have implications for safety. The first relates to the use of flammable fabrics and the second to the design of the clothes worn by some female flight attendants.

The use of nylon and other synthetic fibres for cabin crew uniforms was the subject in 1975 of a notice of proposed rule-making (NPRM) issued by the FAA as a result of concern about the ready ignition of these fibres and their tendency to burn independently after ignition. It is also a feature of these synthetic fibres that they may melt in contact with flames or heat. However, the NPRM was withdrawn in 1981 after the comments received by the FAA indicated that rule-making regarding the flammability of uniforms would not be feasible. It is interesting to note that surveys of flight attendant union members found that the majority of those responding were not "willing to sacrifice style, comfort, or cleanability to obtain uniforms of higher flame resistance" (F03).

The major problem in relation to design is confined to some oriental airlines in which female flight attendants wear national costume (sari, muu-muu, kimono). These attractive outfits may have serious disadvantages in emergencies. Long skirts can impede vigorous movement and may increase the time taken to evacuate in an emergency, particularly from an overwing exit. In one accident in Honolulu, a rejected take-off resulted in an emergency evacuation of the aircraft in which all the flight attendants sustained extensive buttocks friction burns when their loose-fitting muu-muu skirts slipped over their hips exposing their skin to abrasions from the slide (F10).

CURRENT TRENDS

There have been many changes in passenger transport aviation since the early post-war years. Technological advances have resulted in larger, faster aircraft and the passengers are not confined to the upper echelons of society. Changes in the flight attendant's job have resulted from these technological changes and from changes in the economic climate, the most extreme example of which is to be found in the United States as a consequence of deregulation.

Compared with the post-war period when only relatively wealthy people travelled by air, nowadays passengers are drawn from a wider range of socio-economic groups; the aircraft are very much larger; the sectors flown by the aircraft are longer and the stops between sectors are shorter. In spite of the increased capacity for passengers, the increase in cabin staff to administer to these passengers has not always been proportionate.

There are more female than male flight attendants. The

emphasis placed upon the requirements for flight attendants to conform to a particular size and appearance and the need for high standards of personal grooming lead to the perception that the job has a certain glamour. The appeal is enhanced by the opportunity for travel both as part of the job and at reduced cost for private travel.

A survey in the 1970s of flight attendants from 12 airlines in the United States showed a high level of job satisfaction. This was derived mainly from two factors; namely the amount of free time allowed in the job and the opportunity for helping passengers with their needs (R04). It is likely, however, that such a survey in the 1980s would not confirm these findings. Deregulation of the airlines in the United States is reported to have had a deleterious effect on working conditions and "it would be hard to find [a flight attendant] to describe the job as either glamorous ... or as well paid with reasonable conditions as it was in the recent past" (M16).

Deregulation has led to intense competition for customers. At the lower end of the market, airlines compete by reducing the price of tickets. For business travellers and others "up-market", competition centres on the range of the services offered in flight. Both of these market strategies have implications for cabin crew with the result that in spite of the many technological developments since the end of the Second World War, the job of flight attendant is as arduous as ever but without many of the accompanying satisfactions of earlier times.

Lower prices of airline tickets can be achieved by economies in the quality of the service provided. The cut-price fares attract those who might not otherwise consider air travel and the resulting change in the pool of potential passengers, the large numbers of passengers in the aircraft and the reduced level of service offered all have an impact on the way the cabin crew do their jobs and the satisfactions which are derived from them. In addition, the discrepancy between the advertisements which display smiling flight attendants implicitly promising personal service in comfortable, under-occupied aircraft and the reality of reduced seat pitch, crowded aircraft, and minimal service leads to feelings of resentment at being misled on the part of passengers. This, in turn, may be expressed to flight attendants in ways which are distressing.

Competition between airlines for the business executive or the wealthier traveller centres in part on reducing the duration of travel times but also on the level of the services offered and on the charm, friendliness and caring qualities of the cabin staff. The implications of this for flight attendants, particularly on short-haul flights, is that the period of the flight is one of intense activity and there is little time between the programmed round of drinks, snacks, meals, and tax-free goods to give personal attention to any

individuals who might require some particular service.   There is
also a considerable onus on flight attendants to appear constantly
charming, friendly, and caring regardless of the stress involved in
maintaining tight schedules, in placating dissatisfied passengers and
in controlling unruly ones.

Competition leads also to the requirement for full utilization of
both aircraft and personnel.   Scheduling takes little account of the
internal human clock and minimal time is available for adjustment
to time zone changes.

One serious consequence of deregulation in the US has been the
two-tier salary structure in airlines in which those who are engaged
after 1983 are paid at a much lower rate than those engaged before
this date and the associated benefits such as pension funds are
correspondingly reduced.

All these changes lead to conditions of work which are unlikely
to favour the recruitment of people with a commitment to a long-
term career in aviation.   However, in the late 1980s in the United
Kingdom, there appeared to be an excess of applicants for jobs as
flight attendants over the positions available.   In one UK airline,
there were 40 applications for every job vacancy.   In such a
climate, it is possible to be more selective in recruitment.   British
Airways advertisements in the summer of 1988 indicated that the
company was making a bid to upgrade the quality of their recruits
by seeking graduates with foreign language skills for cabin staff.
The possibility of career progression into management was mentioned.   Loss of trained staff is reduced by the policy of permitting
women to take charge of the cabin and to occupy senior supervisory
posts.

The desirability of the job of flight attendant in the Asian
countries is evidenced by the ratio of applicants to vacancies, which
may be as much as 100 to one and by the high quality of the
applicants.   The early compulsory retirement age of 35 or 40 for
women (later for men) in some Asian airlines does not appear to be
a deterrent, nor is it likely that demographic trends will alter the
dimensions of this labour market.

However, with the predicted short-fall in young people in Europe
and the West during the 1990s due to a falling birth-rate, there is
likely to be enormous competition among employers for those young
people who are available and suitable.   This, together with the
growth in air transport and thus an increased demand for staff, is
likely to change the position for airlines in relation to recruitment
from one of plenty to one of shortage.   In this context, there is
every reason for an airline to consider carefully the tasks which
make up the job of flight attendant, the rewards, the conditions of
work, the career structure, and the opportunities for part-time
employment.   Without such detailed attention to the management of
human resources, it may prove impossible to attract and retain

personnel of the appropriate calibre.

REVIEW

Cabin staff are selected and trained to equip them for the two
facets of their task, the provision of normal cabin services and the
preparedness to deal with emergencies. The latter aspect, whilst
primary in terms of the requirements of the regulatory authorities,
is secondary in an analysis of the actual job content.

Periodic training and checking are permanent features of the
cabin attendant's job, whether or not new variations of aircraft are
to be flown. Certain aspects, such as the effects of sleep disturbance, introduce an element of stress into the job, and cause some
crew members to leave their occupation.

Changes in the nature of air transportation, including the speed of
travel and the reduction in the size of flight-deck crews, has led to
substantial increases in the responsibilities of flight attendants.
Consequently many more men and women are required to consider
the cabin as providing a long-term career.

6 The Scope of
Human Factors

Human Factors addresses those issues which arise from the inclusion of Liveware within a SHEL system. The various aspects of the technology are described and attempts made to dispel some widespread misunderstandings.

LIVEWARE

It is clear from a consideration of the aircraft cabin, its passengers, and its crew that a good deal of information about human beings is necessary if the total system is to be designed and managed such that the maximum degrees of safety, of efficiency, and of comfort are to be achieved.

The SHEL model, outlined in Chapter 1, provides one way of introducing a systematic classification of the complex issues which arise. The "three-dimensional" diagram is illustrated in Figure 6.1. If each type of resource - Hardware (H), Software (S), and Liveware (L) - is to interact successfully with each other component, then the interfaces, or points of interaction, must be carefully designed to avoid the friction which can result in inefficiency, delays, or even complete breakdown. Human Factors (HF) is concerned to utilize data about human beings which are relevant to the design and operation of a particular system.

The idea of expertise in mechanical engineering, in electronics, or in the law is commonplace. There is, however, a widespread reluctance to accept that human performance is something which requires the same type of detailed study. It is often said that knowledge about people is simply "common sense". But it is easy to demonstrate, and experience of system operation and failure clearly shows, that common sense is grossly deficient in providing the data needed to design successfully the complex systems which form an integral part of our twentieth-century lives.

HF, then, provides the expertise which is necessary to deal with the problems arising from the inclusion of Liveware within systems. This technology occupies a place of particular significance as a consequence of the relative invariance of human characteristics. In designing a door which is to be operated manually, for example, the initial design parameters should be those derived from human performance data. Problems will result if the engineering of the interface is conducted in the reverse sequence; human beings cannot be redesigned to cope with a task which exceeds their natural capabilities. The general principle, then, is that human characteristics must be accepted as "given", and both Hardware-Liveware interfaces and Software-Liveware interfaces must be engineered within the bounds of human limitations.

Only two techniques which operate counter to this rule are permissible. The first of these is personnel selection. Once job demands are defined, steps may be taken to recruit the most suitable people and eliminate those less able, or completely unable, to achieve an adequate standard. The second technique is training, a process of fitting the Liveware to the job by means of schedules of behaviour modification. Whilst it will be recognised that each of these activities has an important place in system design, it should also be noted that excessive demands embedded within either selection or training programmes may indicate that inadequate attention has been paid to alternative techniques in design. Good system practice demands that proper attention is paid to the loads imposed.

The question has often been raised whether there is a need to coin such a word as "Liveware" when other terms are readily available. It has even been thought to be demeaning for a person to be so described. Such a view derives from a misunderstanding, since the reverse is the case. People are conceived as Liveware components when it is necessary to focus upon their roles within particular systems. It is fully recognised that such roles fall far short of the full description of any unique human being who will have interests, ambitions, relationships, qualities, and achievements which are completely unrelated to the individual's contribution to any one system. A cabin attendant, for example, will typically have a family and friends, religious and political views, leisure and cultural interests. The total person is the summation of all these things. The system designer, whilst recognising the unique nature of each human being, must attend to that sub-set of characteristics which is of relevance to a particular task. In deciding, for example, whether a given function is to be performed automatically by a Hardware "black box" or by a Liveware component, the designer considers only the relevant properties of each. It is in this sense that "Liveware" is different in meaning from "people".

Related to the above discussion is the second reason why system design must place human considerations in a primary position.

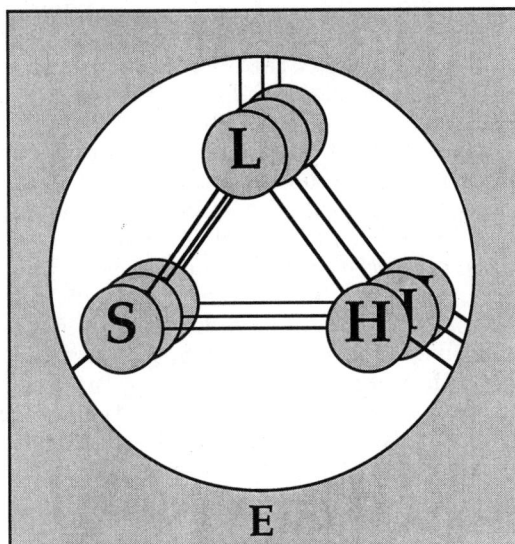

Figure 6.1 The SHEL model. In a well-designed and managed system the components interact in a harmonious way. A disturbance at any point might bring about an imbalance requiring numerous readjustments elsewhere in the system.

Unlike other system components, people have intrinsic value. A passenger aircraft, removed from its function as a means of transport, is simply a piece of scrap metal, albeit an expensively produced one. Similarly, an item of Software, such as a Standard Operating Procedure, has no value if found to be inadequate, or no longer applicable to the prevailing conditions. It can be discarded and all references to it expunged from the manuals. There can be, however, no scrap heap for people. Certainly Liveware can be "automated out" of particular systems, but the value of those people who once performed such redundant functions remains intact.

HUMAN ERROR

It has been said by many people throughout the centuries that error is an essential ingredient of the human condition. The evidence favouring such a view is undeniable. The issue to be addressed is not the truth or falsity of the claim, but that of determining strategies which should follow from its acceptance.

Numerous doctrines of despair have been advanced from time to time. It has been said, for example, that human error is simply something that must be accepted as inevitable; that there is no point in complaining about it; that we must learn to live with it. Another reaction is that human beings, since they are error-prone, should be removed from situations in which high penalties can result from error and that human performance should be replaced by mechanical or electrical equipment.

No such doctrine is acceptable within a rational approach to the problem of error. The former attitude completely disregards the various steps which can be taken to control both the frequency and the consequences of error. The latter approach naively overlooks the fact that those who design, manage, maintain, and modify automatic systems are no less susceptible to error than those who operate them. Furthermore, the philosophy of "throwing out the baby with the bathwater" can introduce penalties of its own.

An early stage in the study of any set of phenomena involves the classification of samples of its appearance. So it is with error; numerous ways of classifying errors have been developed and each one provides some insight into the nature of the processes concerned. One such classification includes determining the point within the Input-Processing-Output chain at which things go wrong. Was, for example, a spoken message entirely missed, and if so, was this due to its being insufficiently loud? If the signal was not overlooked, was this due to some ambiguity in the message, or to the presence of background noise? Was the correct response intended but the action fumbled, or was there an error in deciding how to respond? Very many such questions may be posed, the answers to which provide an understanding of the ways in which information may fail to produce the appropriate output during the progress through the various stages of the information chain.

A second example of an error classification technique might be concerned to seek patterns in the type of error most commonly encountered. Do people tend regularly to err in a particular direction, or is there no sign of such consistency? Are the errors spread fairly uniformly over the available alternatives or are there one or two commonly recurring erroneous selections? In the event that a consistent pattern emerges, then specific steps might be taken to modify the form of the information input, or to adopt such other methods as training schedules to ensure that the bias is removed. Alternatively, if it is found that performance normally achieves a high standard but occasional gross errors are observed, it will be wise to search for a particular disturbing influence such as distraction or fatigue.

The classification of errors in these ways, such that their nature and the circumstances surrounding their commission may be described, provides the background data upon which error-reducing techniques are based. The appropriate steps to be taken will vary in different circumstances but could include modifications to equipment ie improving the H-L interface; changes in standard procedure, ie improving the S-L interface; revisions of training schedules, ie modifying L; development of communication skills, ie improving the L-L interface. It will be necessary to evaluate each of a number of such possible remedies before deciding upon a particular error-reducing strategy to be adopted.

Should residual errors be allowed to lead to unfortunate consequences? The clear negative reply to such a question is the basis for the second phase of the HF programme to combat the effects of error. This comprises the introduction of methods which will ensure that any remaining errors are detected, identified, and corrected. Under normal operating circumstances, feedback loops can quite easily be established to ensure that such processes are in place. On using the cabin address equipment, for example, a flight attendant will ensure that the message is being broadcast throughout the cabin. Should there be any doubt about this, then it may be necessary to enlist the assistance of a colleague or a passenger. Every critical action should be checked. On occasions, Hardware devices such as warning lights may be available to indicate the effects of action: at other times human observers should monitor the performance of their colleagues.

Each individual cabin attendant, and particularly those involved in management and in training, would be well advised to analyze the possible effects of any errors which could be made. In all those cases where the consequences are non-trivial, some thought should be given to the mechanism by which such errors would be detected and subsequently corrected. An accident has been defined as an error with unfortunate consequences. In the event, therefore, that no error-detecting mechanism exists, then an accident may be waiting to happen. A flaw in the system has been revealed and steps should be taken to ensure that adequate corrective measures are established.

Aircraft accident investigators determine the major causes to which each disaster is attributed. It is well-known that reviews of accident reports consistently show that human error is the category most frequently used, appearing in about two out of every three accidents. Clearly, then, flight safety is largely a matter of error control which, in turn, is dependent upon the efficacy of the techniques employed in the design and management of SHEL systems. It is unfortunate that Liveware remains the least adequately accommodated system resource, due to the relatively small attention paid to the applications of HF.

LIVEWARE: THE SYSTEM COMPONENT

The Liveware characteristics to be considered within the context of planning the construction and operation of a system can be conveniently summarised under six headings as shown in Table 6.1. In this way, information is presented such that it maintains some comparability with engineering data books.

The amount of information to be presented within each section will vary according to the needs of the user. Whilst a great deal is already known about human beings, the life sciences remain areas of research in which many problems remain unsolved.

Physical size and shape

Anthropometry is the study of human body dimensions (N02). Applications are to be found in the production of clothing in different sizes, in the design of buildings, vehicles, furniture; in fact in all spaces or equipment to be used by people. Stature, or height, is probably the most widely used dimension and a person's height provides some guidance about the probable size of arm reach, a leg length, and other bodily sizes. However, people vary in shape; and predictions based upon stature alone can be unreliable.

Table 6.1 Liveware characteristics can be conveniently summarised under six headings. Such a list might be regarded as the Contents page of a Liveware Data Manual

---

                    Liveware Data

1.	Physical size and shape
2.	Power output limits
3.	Information input properties
4.	Central processing capacities
5.	Information output properties
6.	Limitations of ambient conditions

The mean, or average, value of a dimension is often used as the best compromise in design, but it is important too for a designer to be aware of the range of sizes to be accommodated. In locating a high shelf, for example, the upward reach of a small woman is the essential criterion, whereas in designing a doorway the stature of a tall man is the critical value. In some cases, when a precise relationship is necessary between Hardware and Liveware, an adjustable range must be provided to accommodate the population of users. An example is to be found in the pilots' seat on the aircraft flight-deck.

Differences in body size derive from the age, sex, and racial origin of groups of people in addition to the variability to be found within any one group. Examples of such differences are to be found in Table 6.2. Clearly it is important for designers to be aware of such differences. Many published sets of data provide information upon a large number of body measurements.

Table 6.2 Some examples of anthropometric variability: values of three dimensions from each of three different groups. It will be seen that men are larger than women, and European men are larger than Asians. Asians' shape is such that the trunk is comparatively long and the limbs short.

Dimensions (mm)	Males		Females
	Asian	European	
Stature	1660	1750	1640
Seated eye-height	770	800	750
Buttock-knee length	540	605	570

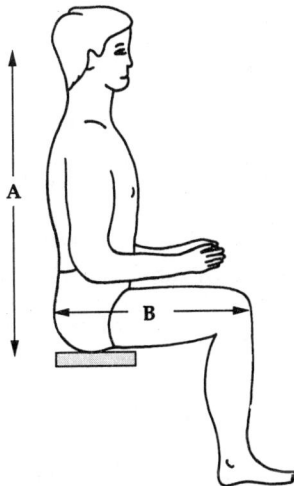

Figure 6.2 Two of the body dimensions described in Table 6.2. (A = eye height, sitting erect. B = buttock-knee length.)

Static measurements taken in precisely-defined rigid postures provide only part of the required data (K01). In specifying good seating, for example, it is necessary to know how the body should be supported in order to encourage postures which minimize fatigue and long-term damage to the spine and musculature (K08). Dynamic anthropometry is concerned to study the space requirements necessary for the performance of such activities as climbing a staircase, reaching under a seat for a life-vest, or exiting swiftly along an aircraft aisle. Numerous photographic and other techniques have been developed in order to record such three-dimensional data (G18).

Power Output

A Liveware component displays some similarity to other heat engines. There are fuel requirements, operating limits, and working efficiencies to be considered (W07). Since human energy is produced by the oxidation of substances derived from food, the important variables are the quantity and quality of diet and the availability of sufficient oxygen to release energy at the time it is required.

One important difference between Liveware and physical engines should be noted. It is possible to specify the power output of a Hardware component simply in terms of horsepower or watts.

Liveware power may be expressed in those units only if the duration of the output is also specified. Short periods of high power output far in excess of levels maintainable for a complete working day must be followed by rest pauses to allow for recovery.

The load imposed upon passengers is normally very low. Cabin crew, however, may be called upon to sustain quite high outputs during certain activities in the galley, pushing trolleys, and conducting other service functions. In the event of an emergency evacuation, large amounts of energy will be used by both crew and passengers in effecting their escape.

It is possible to estimate the power required in the performance of tasks during the process of designing equipment and procedures. Such estimates should be carried out taking due account of the age, sex, and racial origin of the personnel forming the potential user group.

Whilst the high-energy compounds produced by the body from food can be stored for long periods, the oxygen necessary to allow the release of energy must be more readily available. In a restful sitting position, a person may require about 0.35 l/min of oxygen. The requirement during the performance of tasks in the cabin might be two or three times this value. The most hazardous situation is therefore one in which the oxygen supply is curtailed. In flights above 10,000ft such conditions may prevail in the event of a failure in the pressurization system or a breach in the pressure hull. Although muscular activity can continue for relatively long periods, resulting in oxygen debts - as in athletic performance - the brain demands a continuous oxygen supply. Deprivations of a few seconds, particularly if combined with muscular work, can lead to loss of consciousness, permanent brain damage, and death.

Information Input

It is normally said that man has five senses, namely sight, hearing, touch, taste, and smell. In fact, there are several additional channels concerned with the sensation of temperature, pressure, pain, orientation, and motion. Each of these plays an important role in the defence and well-being of the individual.

When designing ways of providing information to people the visual and auditory channels are normally selected due to their versatility and high capacity. For inputs to be reliably and economically perceived, the messages must be suited to the human sensory and auditory mechanisms. Visual signals, for example, must conform to standards of size, brightness, and contrast; auditory signals to standards of loudness and pitch.

Over the last forty years, a huge literature has accumulated describing how human performance is affected by the way in which information is presented. This part of HF, normally termed "Display Design", comprises one of its largest subject areas. The available knowledge concerning visual inputs is applicable not only to mechanical and electronic instruments but also to printed text and illustrations, placards, and signs. Speech, of course, provides the primary form of auditory input, but many systems make use also of coded sound signals of various sorts.

Central Processing

Signals which succeed in conforming to the requirements of sensory channels will not necessarily be effectively handled by the central processing mechanisms since here there are further highly complex processes in operation. The sensory channels may, for example, be in competition with one another. The gap in a conversation with a motor-vehicle passenger will be familiar to the driver who has just completed a particularly difficult manoeuvre. Human information processing has been described as a "single channel" process in which items must be handled one at a time.

Simultaneous inputs are by no means the only problem. The reception of incoming information will also be affected by our expectation and even our wishes. A recurring feature of certain types of errors and of accidents is the persistence of the "false hypothesis". We tend to enter situations with certain expectations about what we might perceive, and to distort any contrary signals such that expectations are fulfilled. Undue weight is given to any evidence which apparently supports the hypothesis, and conflicts are ignored. This phenomenon is more pronounced in the presence of stress.

Processing is facilitated by the existence of short-term memory which preserves information in a buffer for a short interval before it is lost. In addition, we possess a high-capacity long-term memory, retrieval from which is governed by highly complex associative links.

To some extent, the human capacity to process information and thereby to comprehend messages, solve problems, carry out calculations, and make rational decisions can be described in logical terms such that the human facility may be compared with Hardware computing devices. There is, however, a danger of over-simplifying the description by overlooking the associative and emotional aspects of human reasoning. The influence of beliefs, fears, wishes, hopes, prejudices, and expectations cannot be ignored.

Output Properties
In addition to the forces people can exert in order to lift, carry, and manipulate objects, outputs can be described in terms of their information content. Different parts of the body are better adapted than others for performing swift and accurate movements having high information content. The fingers, for example, can perform more skilful tasks than can the legs and feet.

Control movements are made up of a series of discrete elements each one of which is planned, albeit unconsciously, prior to its execution. As any pianist, typist, or snooker-player is aware, a great deal of practice may be necessary before such movements can be reliably and accurately produced. Even an exercise involving almost every muscle in the body, such as the golf swing, is a pre-programmed ballistic movement, in spite of many players' subjective impression to the contrary.

The most remarkable feats of coordinated muscular activity are those involved in the production of speech. Very large amounts of information are processed and transmitted to the muscles which control the movements of the lips, jaw, tongue, and throat to produce a wide range of swiftly changing sound effects.

Limitations of ambient conditions
The final item on the Liveware Data Sheet consists of descriptions of the ways in which human performance is affected by features of the ambient surroundings. Three such aspects, illumination, noise, and the thermal environment, are relevant to almost any situation in which performance can be observed. Less common hazards

include vibration, air pollution, and radiation. In each case, data are available describing the effects of these, and recommendations have been published concerning optimal and deleterious levels (N03).

In the contemporary cabins of jet aircraft, most hazards are kept under control such that risks of resulting damage to crew and passengers are extremely low. There remain, however, areas in which some annoyance may result from noise and vibration, and a certain amount of discomfort from low humidity and air pollution.

Some of these problems persist as a result of the economic penalties associated with their solution. Humidity serves as an example. Very large quantities of water would be necessary in order to maintain a comfortable level throughout a long flight at high altitude. Other problems arise from conflicting requirements of various cabin occupants. When some passengers wish to have darkness, warmth, and quiet in order to sleep, others will require conditions matched to their wakeful activities. Many of the most acrimonious clashes occur in relation to smoking, as a result of the failure to segregate smokers from those who prefer a completely smoke-free atmosphere.

THE INFORMATION LOOP

The input, central, and output processes together form an integrated chain whereby incoming signals bring about an appropriate response. In the performance of many skilled tasks, this chain forms part of a loop involving elements in the outside world.

The aircraft pilot, for example, receives information from his instruments, from his outside view, by radio from Air Traffic Control (ATC), and elsewhere. This information is fitted into a framework which has been built into his memory by training and by experience. The appropriate outputs, in the form of control movements, verbal commands, or other forms of behaviour, are then formulated and implemented. Such outputs will, in many instances, contribute to the succeeding inputs, thus closing the loop.

Analogous information loops describe the behaviour of cabin crews, who will perceive certain states relating to their duties and employ their skills to evoke the relevant responses. The results of their actions will contribute to succeeding inputs. The design of interfaces involving human performance is thus concerned with determining ways in which the incoming information may best be presented to ensure that it is promptly and accurately perceived, and that the consequent responses are facilitated in ways which minimise errors, delays, or difficulties. A large part of any training programme is devoted to building up and rehearsing the human portion of information loops by teaching people to perceive and interpret certain types of input signals and to utilize the skills they are acquiring in order to produce the optimal responses.

Communications serve as an example of information loops which are of critical importance within the cabin. The attendant will perceive a message originating from a passenger. This may

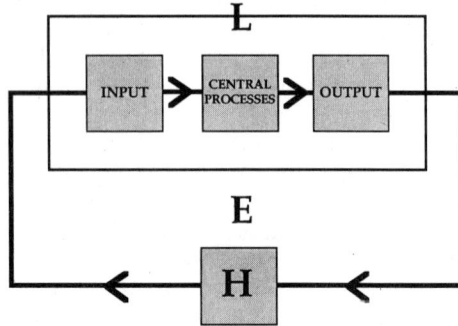

Figure 6.3 The information loop involving Liveware and machinery.

comprise a simple spoken question or may be a more complex event from which the observant attendant gains additional information from the passenger's posture, facial expression, or speech style. The response offered will provide analogous sources of information to the passenger such that an information loop continues to operate throughout the course of a conversation. Communication loops may, of course, involve more than two people, so that information flows in several directions simultaneously.

Since communication forms part of everyday life, and is one of the social skills acquired from contacts with family, friends and colleagues, there is a tendency to believe that no special training is required in order to become an effective practitioner. Such a view is unsound. Within the cabin, highly developed skills are required in order to communicate effectively with the variety of passengers to be encountered, and with colleagues in the cabin and on the flight-deck. Most especially in the case of emergencies are demands made upon the attendants' ability to convey information, advice, and commands in the most effective manner.

INFORMATION DESIGN

Analogous HF principles apply to alternative modes of communication such as those involving printed text, tables and graphs, pictures, diagrams, placards, maps and charts, signs, or notices (W10). In each case, the information source must be implemented using appropriate physical values, matched to human input mechanisms. This is probably the most easily achieved phase

in the design process.   Attention must also be paid to those aspects
which relate to the manner in which human information processing
takes place.   In this phase, the relevant features include the shape
and format of the information, the ways in which meaning is
represented, the sequence of presentation, the principles of
classification, the rules of coding, and the provision for the user to
navigate within the information.

The questions which arise during the consideration of these issues
will take account of the circumstances in which the information is
to be used, by whom, and - most particularly - for what purpose.

An emergency briefing card for passengers, for example, presents
a different set of design requirements from that of a crew training
manual. Each needs to be clear, complete and unambiguous, but
their purpose and circumstances of use differ widely.   The card
embodies a set of instructions without explanation of their origin;
its content must be grasped swiftly by untrained personnel; its dependence upon any particular language must be slight.   A training
manual, on the other hand, is intended for use during a protracted
period of study; its various sections may be consulted repeatedly
without undue time stress; the reader will be expected to have
some grasp of the rationale underlying the instructions it contains;
users will be expected to have a good command of the language in
use, together with certain preliminary educational standards.

Whilst the similarities between the two example documents will
dictate a degree of commonality in the principles of information
presentation, their dissimilarities will dictate different solutions to
the design problems posed.

LIVEWARE INTERFACES

HF begins with a description of the characteristics, capabilities, and
limitations of Liveware.   Once a task within a system has been
defined, it becomes possible to consider the suitable personnel
selection criteria which might be employed, and to evolve a training
programme.   Each of these activities can call upon the experience
gained during the long history of applied psychology.

As a system is being developed by means of the specification and
procurement of the various resources, the primary role of HF lies in
designing the interface between Liveware and other system
components. In Figure 6.4, those parts of a SHEL system which
directly involve Liveware are illustrated.   This anthropocentric
diagram provides a simplified way of depicting the four interfaces
of which L is a component.   It must, of course, be remembered
that other aspects of the SHEL system remain unrepresented in this
diagram.

Experience has shown that most difficulties and disasters which occur in operational circumstances are due not to the catastrophic failure of single components, but to mismatches at the interfaces. Thus, whilst it is possible for a crew member to suffer sudden incapacitation, or for an item of Hardware to fail with disastrous

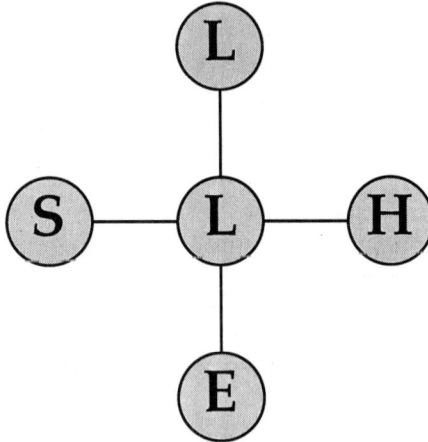

Figure 6.4 The anthropocentric diagram illustrates those parts of the SHEL model in which an interface with Liveware is included.

results, it is far more usual for errors and accidents to arise from an instrument misreading (L-H interface), the failure to understand the relevant procedure (L-S interface), or from a breakdown in communication (L-L interface).

In the chapters which follow, the HF aspects of cabin emergencies are discussed in terms of the type of system component primarily involved.

PART II
Emergencies

7 The Major Hazards

It is, of course, necessary to be aware of the nature of potential hazards before attempting to examine the HF contribution to the reduction of their incidence and the containment of their effects. An aircraft operates in a hostile environment, and is most vulnerable during the course of leaving and returning to the ground. However, with the exception of light turbulence and very minor fires, the hazards reviewed here are only very rarely experienced.

FLIGHT SAFETY

In view of the large number of organizations, publications, and conferences devoted to the furtherance of flight safety, it might be supposed that aviation provides a particularly hazardous mode of transport. Quite the opposite is true. Whatever yardstick is chosen to assess the comparative risks involved, public air transport is shown to be one of the safest methods of journeying from one place to another.

It is important to understand something of the public perception of risk and danger in order to place in context the issues relating to flight safety. Essentially, the rule is that if many people are killed or injured simultaneously in one place, then the media and the public react with vigour to the perceived danger. Alternatively, if several scattered events occur in which only one or two people are involved then such events, despite having a common causal background, are likely to be ignored. Accidents to large public transport aircraft do, on occasion, involve heavy casualties and attract public attention. They will be followed by intensive investigations and, possibly, by public enquiries. The ultimate published conclusions of these activities will attract further attention and increase the perceived risk associated with flying.

This general rule may be illustrated by reference to such events as a tragic episode in Britain in 1987 where 16 people died as a result of the use of firearms. Widespread public debate followed substantial media coverage, including many interviews with firearms experts and politicians. There was immediate discussion about law reform in an attempt to prevent reccurrence. Changes in the law followed in due course. Within 24 hours of the tragedy, the Prime Minister arranged to inspect the scene and to meet some of the injured and bereaved. Yet in the interval between the event and her arrival, more people in the United Kingdom had died as a consequence of road accidents than had been killed by the gunfire. These scattered tragedies on the roads led to no news headlines, no interviews with road safety experts, no particular public reaction, no changes in the law.

In recent years, flight safety has improved to the point at which a fatal accident will occur on public transport world-wide at the rate of about once for every 600,000 flights. These are extremely long odds. Statistics from developed countries show that by far the largest numbers of fatal accidents occur either in the home or on the roads.

A further widely held misconception is that aircraft accidents typically result in the catastrophic loss of life for everyone on board. Such is not the case. More than 80% of passengers survive accidents without sustaining injury.

An important difference between aviation and road accidents is the relative occurrence of injury and death. Comparisons only of fatalities can yield a misleading picture. Whilst the number of serious injuries on the roads far exceeds the number of deaths, the reverse is true in the case of aircraft accidents.

How, then, have the present-day levels of safety in the air been achieved in spite of the potential dangers of operating in a hostile environment away from the surface of the earth? Over the years, much effort has been expended to establish mechanisms for the achievement and maintenance of acceptable standards of operation of air transport. Each country has established an Authority which bears responsibility for setting and policing standards within the industry. These standards include the certification of aircraft, the licensing of crews and maintenance personnel, and the regulation of procedures and conditions of operation including such aspects as maximum periods of duty. There are high levels of international communication and cooperation through the International Civil Aviation Organization (ICAO). Many other bodies exist at national, regional, and international levels, striving to find ways in which flight safety might be enhanced.

Considerable resources are expended in order to achieve high flight safety standards. It could be argued that more benefit would result to society as a whole were some of these resources diverted to other areas. In this way, it seems certain that total levels of fatality, injury, and damage to property would be reduced. Whether such a policy would gain wide public support is quite another matter. Certainly no support for such a policy would be forthcoming from the aviation industry.

It is widely accepted that complacency in matters of aviation safety provides a potential source of deterioration in standards. For this reason, and in spite of the overall safety record, each source of hazard must be carefully examined with the objective of eliminating, or at least reducing to a minimum, its potential threat.

Before discussing these major hazards individually, it may be useful to note some definitions. An aircraft accident as defined in Annex 13 of the ICAO Convention is

"an occurrence associated with the operation of an aircraft which takes place between the time any person boards the aircraft with the intention of flight until such time as all such persons have disembarked in which a) a person is fatally or seriously injured as a result of being in or upon the aircraft, or by direct contact with any part of the aircraft, including parts which have become detached from the aircraft, or direct exposure to jet blast; or b) the aircraft incurs damage or structural failure which adversely affects the structure strength, performance or flight characteristics of the aircraft and which would normally require major repair or replacement of the affected component; or c) the aircraft is missing or is completely inaccessible".

An incident is defined as
"an occurrence, other than an accident, associated with the operation of an aircraft which affects or could affect the safety of the operation".

The universal acceptance of these definitions makes possible the collection and collation of data on a world-wide basis. There may also be legal significance in the consequences associated with different types of event. The definition of a survivable accident, on the other hand, is in a different category. It has no legal status but is useful in delineating a class of events the investigation of which is likely to prove helpful in accident reduction. The definition of a survivable accident used by the US National Transportation Safety Board (NTSB) states that it is

"an accident in which the forces transmitted to the occupant through his seat and restraint system do not exceed the limits of human tolerance to abrupt accelerations and in which the structure in the occupant's immediate environment remains substantially intact to the extent that a livable volume is provided for the occupants throughout the crash sequence" (N17).

A partially survivable accident is one in which the crash forces in part of the aircraft are non-survivable while other parts are within the survival envelope (N17). Elsewhere, a survivable accident has been equated with a non-fatal accident, and a fatal accident defined as one in which there are no survivors (M30).

Six major types of hazard are discussed below. Turbulence, decompression, and fire are hazards which may be encountered during flight at altitude. Other major hazards comprise abnormal contact with the ground, or with water, and the fires which may develop following a crash. Additionally, and in a different category, are the hazards associated with terrorism and the destruction of aircraft by explosive devices. These will be touched upon briefly.

TURBULENCE

The captain of a Douglas DC-8, in reporting his encounter with turbulence described how he "....experienced violent buffeting. Briefcases, manuals, pencils, cigarettes and flash lights were flying about the cockpit like unguided missiles. It sounded as if....the structure was disintegrating. The objects that were thrashing about the cockpit seemed momentarily to settle on the ceiling....I had the feeling we were inverted as my seat belt was tight and had stretched considerably" (F13).

Turbulence is the commonest cause of in-flight accidents. One source of this vertical air movement is the thermal convection resulting in the production of cloud formations. More insidious are the phenomena leading to clear air turbulence (CAT) since these are less easily identified and avoided. In terms of acceleration forces, turbulence is described as moderate between 0.5G and 1.5G and severe above 1.5G. Turbulence may also be classified in terms of its effect on unsecured objects, the service of food, the ability to walk about, and the subjective feelings of strain upon a seat belt (see Table 7.1).

The forces experienced in the rear of the aircraft may be up to 50% greater than those experienced in the front. This is because there is a rotational component in the turbulence; hence its effects are distributed in proportion to the displacement from the axis of rotation.

An aircraft may encounter light turbulence over a long period of time, extending sometimes for the duration of the flight. Moderate to severe bursts of turbulence can last for many minutes.

Incidence
It has been estimated that once every hour throughout the year there is a CAT encounter over the United States. For those of

Table 7.1   Classification of turbulence

Light	Slight displacement of unsecured objects food service may be conducted little or no difficulty experienced in walking slight feeling of strain against seat belt
Moderate	unsecured objects dislodged food service difficult walking difficult definite feeling of strain against seat belts
Severe	unsecured objects tossed about food service impossible walking impossible occupants forced against seat belts

moderate intensity and greater, the estimate is two every day.

Cabin crews may be expected to meet moderate CAT once every six months.

Between 1976 and 1987, around 10 accidents a year to public transport aircraft were attributed to turbulence. These accounted for nearly 10% of the total (E04) and caused serious injuries to 88 crew and 248 passengers.

Table 7.2 Injuries from turbulence, US, 1968-71 (N06)

Passenger		Flight Attendant	
Serious	Minor	Serious	Minor
43	91	63	60

Injury
The injuries incurred in turbulence are caused by loose articles flying about and hitting people, and by people being thrown about and contacting hard or sharp surfaces. The danger of being thrown about may be avoided if passengers are seated with seat belts fastened. However, overhead lockers have burst open during turbulence and the contents, which may include heavy objects such as bottles of tax-free alcohol and briefcases, have been thrown

about causing injury to those seated beneath.

Turbulence described as "light" by flight-deck crew has caused injury to individuals at the rear of the aircraft.   Moderate levels of turbulence have caused unconsciousness and serious injury to passengers and flight attendants. These have included fractures to limbs and ribs; injuries to the head, neck and back; and lacerations and contusions. In one incident, four passengers and one flight attendant who were in the rear of an aircraft during moderate turbulence were so badly injured that they had to be hospitalized (M04).

Very high forces (up to 2.5G) have been experienced during episodes of severe turbulence. In one incident, flight attendants who were not seated with fastened seat belts were thrown into the air and three passenger service carts hit the ceiling. Four serious and twenty minor injuries were reported (N06).

Differential risk
Since flight attendants are more likely than passengers to be standing or walking about the aircraft or working in the galley, their exposure to risk of injury is greater. Evidence of this effect can be seen in a study based on accident data from 1976 to 1987 which showed that flight attendants were more than four times more likely than passengers to incur serious injury due to turbulence (E04). Injuries to flight-deck crew as a consequence of turbulence are extremely rare.

Because the greatest vertical movement is experienced in the rear of the aircraft, the highest risk is to occupants there who are standing. On one occasion, a passenger standing at the rear of the aircraft when the seat belt sign was illuminated was thrown into the air and broke her leg on falling to the floor (A01). For flight attendants in the rearmost galley, where they are unlikely to be seated, the effects of location are combined with the hazards of unsecured objects which are often intrinsically dangerous, such as hot coffee-pots and heavy service carts. Service carts can become hazardous missiles if not properly secured.

Injury reduction
The way to reduce these turbulence-induced injuries may be easier to specify than to implement. Stronger locking devices on overhead lockers, more effective means of anchoring service carts, and better design of food preparation areas would help to reduce injury. The use of overhead lockers solely for such soft items as coats would minimize the eventuality of hard or heavy objects being thrown about. This would have implications for the amount of hand luggage permitted in the cabin, which is an area of some contention.

The conclusions of a NTSB Safety Study nominated as contributory factors to in-flight passenger injury "the inability of cabin attendants to maintain seat belt discipline effectively and the capriciousness exhibited by passengers who refuse to heed warnings of anticipated turbulence" (N06). This inability on the part of flight attendants to exercize effective discipline was considered to be due to the workload involved in on-going duties connected with passenger service.

It is part of flight attendants' briefings at the beginning of a flight to advise passengers to leave their seat belts fastened whilst seated even when the illuminated seat belt sign has been switched off (FAR 121.571). However, it is commonplace to hear the clicking of the belts being unfastened as soon as the sign is no longer visible. This situation is unlikely to change unless passengers become aware of the value of leaving the seat belt fastened. A survey of 35 incidents of turbulence showed that no injury was reported by anyone whose seat belt was snugly fastened (M04). Although there are studies showing that injuries have been incurred by those wearing seat belts, the evidence shows that keeping the seat belt fastened will minimize possible injury.

It is normal practice for the flight-deck crew to warn cabin staff of impending turbulence. If this is overlooked, then there is a risk of cabin staff incurring avoidable injuries. However, warnings are not always possible when unpredictable bursts of moderate or severe turbulence are embedded in a more enduring period of light turbulence. This was the case where six passengers and two flight attendants were hospitalized after serious injury due to severe turbulence, and twenty-one passengers suffered minor injury. Because of light turbulence, the seat belt signs had not been turned off. When severe turbulence was encountered some passengers were not in their seats. Most of the injuries were incurred by passengers whose seat belts were not fastened (N22).

There is a problem in communication here. The longer the seat belt sign is left on in conditions of continuing light turbulence, the less likely it is to have any effect on passenger behaviour unless reinforced from time to time with oral messages. In addition, passengers may feel reassured about moving around the cabin when they see cabin staff continuing to serve food or drinks while the "fasten seat belt" sign is on.

The NTSB study also recommended that flight attendants "cease in-flight service" and take their seats with restraints fastened when turbulence of moderate or greater intensity is expected not only for their own protection but also to communicate appropriate signals to passengers by their response to the warnings of turbulence (N06).

In order to optimize safety, passengers and cabin crew should remain seated with seat belts fastened whenever there is a forecast of possible turbulence. However, there is a practical problem here. On all but flights of the shortest duration, passengers will expect to have some food and beverages; they will need to walk to the toilet compartment and to move about the cabin. Flight attendants must serve the passengers with food and beverages among their many other duties, all of which must be carried out within a fairly tight timescale.

The best compromise to reconcile these conflicting demands is for the passengers to be encouraged to keep their seat belts fastened at all times during light turbulence and to leave their seats only when it is strictly necessary. Flight attendants should be advised to be very cautious if they have to move about when the seat belt sign is on.

Sudden manoeuvres

There has been mounting public concern about the effects of the increasing numbers of aircraft in the skies (particularly in the vicinity of airports and at certain times of the day) on the probability of in-flight collision. It must be expected that as the number of aircraft movements increases, there will be a corresponding increase in the number of occasions upon which the safe space around each aircraft is violated. However, the totals reported monthly or annually must be treated with caution as there is a considerable degree of variation in the distribution of these rare events.

Swift evasive action may be taken on sighting another aircraft in close proximity. In addition, pilots have, on occasions, taken relatively drastic corrective action on realising that altitude clearances have been violated or upon the appearance of an automatic flight-deck warning. These movements are likely to have a similar effect on occupants as those resulting from turbulence. Occupants standing and, to a lesser extent, those seated but without belts, will be most vulnerable to injury.

An evasive manoeuvre by the pilot of a McDonnell Douglas DC-10 to avoid mid-air collision with a Lockheed L-1011 resulted in serious injury for three passengers, and slight injury for ten flight attendants and 11 passengers. The seat belt sign was on and meals and beverages were being served at the time. Flight attendants and service carts were thrown against the cabin ceiling and three passengers whose belts were unfastened were also thrown about. All unrestrained items which were in contact with the ceiling during the period of the negative G forces fell heavily on to the floor, the seats, and the passengers (N08).

DECOMPRESSION

The altitude at which commercial jet aircraft typically fly is far above that which could be tolerated by an unprotected human being.

For this reason, aircraft cabins are pressurized to produce the equivalent of an altitude which rarely exceeds 8,000ft. Changes in atmospheric pressure within the cabin do take place, however, when the aircraft is climbing or descending, and these changes, particularly during descent, can cause pain and discomfort to some passengers (See Chapter Four).

Causes and effects

Decompression results when, for whatever reason, the pressure in the aircraft interior is no longer maintained at around the equivalent of 6,500 to 8,000ft but reduces until it is equal to the ambient pressure outside the aircraft. This can take place gradually as a result of failure of the mechanisms responsible for maintaining the appropriate pressure (door seals, the air conditioning system) or rapidly, a less common occurrence, when the integrity of the cabin has been breached, for example by structural failure of the fuselage or by an explosion. The effects of rapid decompression on the environment of the passenger cabin are immediate and startling; they include high noise levels caused by the air rushing out of the cabin, dense fogging of the atmosphere due to condensation of water vapour, and a sharp drop in temperature. The effects of slow decompression are not dramatic. There are no obvious changes in the cabin environment. Slow decompression is most readily recognised by the deployment of the oxygen masks.

Incidence

The incidence of decompression is low. Over the ten-year period 1974-1983, the total number of incidents reported to the US Civil Aeromedical Institute (CAMI) was 355, an average of around 35 per year, less than half of which were considered significant. The CAMI criteria for significant decompression are that the cabin pressure exceeds 14,000ft, or that the passenger masks are deployed, or that any injury results from the incident. The injuries associated with the reported incidence of depressurization were few in number. These included one fatality in a passenger who had a pre-existing heart problem, three cases of serious injury, 125 reports of ear pain and 11 of serious ear damage. The effects of hypoxia were reported in 17 cases, five of which were associated with loss of consciousness (H05).

Outside the United States, accidents involving rapid decompression have caused damage to the fuselage sufficient for occupants to be blown out of the cabin. In 1980, a Lockheed L-1011 of Saudi Arabian Airlines experienced an explosive decompression while en route from Dharan to Karachi. The decompression resulted from a failure in the landing gear which created a hole in the cabin floor through which two passengers were ejected and killed (P06). In 1988, structural failure of a Boeing 737 while flying over Hawaii resulted in the loss of a section of the fuselage. In the resulting decompression, a flight attendant was blown out of the aircraft and several passengers were seriously injured (F12). The following year, nine passengers were lost when a hole, about 3m by 6m developed in the skin of a Boeing 747 travelling from Honolulu to New Zealand.

The effects on the body of decompression

Under conditions of slow decompression, the gases trapped in various cavities of the body will expand and cause discomfort and pain. Gas trapped in the gut increases in volume and may give rise to reflex muscular contractions of the intestine which are experienced as abdominal cramps. The muscular contractions transfer the gas to the lower bowel from which it is expelled. Gas trapped in the stomach is expelled through the mouth by belching. Gas trapped in the sinuses and under the teeth can cause pain as the gas expands. Where decompression is rapid, the expansion of the gas is rapid and may eventuate in slowing of the heart rate and loss of consciousness.

However, the major effect of decompression on the human body is the reduction in the oxygen available for vital functions. Oxygen is essential for survival. As pressure decreases there is less oxygen per given volume of air and thus less oxygen in each breath. Pulse rate and breathing rate increase to compensate for this reduction in oxygen, and headaches and nausea may be experienced. This condition is known as hypoxia. Total oxygen deprivation produces a swift loss of consciousness and paralysis. This condition, anoxia, will result in death after about two minutes.

The brain and the eye together consume some 20% of the total oxygen requirements of the body. These organs, having no storage capacity, are highly sensitive to oxygen depletion and are therefore the source of the earliest signs of hypoxia. Irreversible damage will be done if the condition persists.

The impact of hypoxia on an individual is dependent upon age, sex, fitness, and state of fatigue. In addition to these individual factors, the effect varies with exposure time and different severities of oxygen depletion, and with the extent to which the person is physically active.

The critical factor is the gas pressure of the oxygen in the lungs, which determines the rate at which oxygen can be transferred to the blood and thereby transported to the bodily organs. In the context of aviation, it is convenient to express this pressure in terms of the altitude above mean sea level at which such pressure would occur in the standard atmosphere. On this basis it is possible to establish some broad descriptions of those effects which are of operational significance.

Decrements in visual performance, such as some loss of night vision, will be measurable above 3,000ft. At about 8,000ft reaction times will be noticeably longer and mental performance generally will be slower. Supplementary oxygen is necessary for continuous flight above 10,000ft; this has long been considered the maximum safe altitude without respiratory support. Above this altitude progressive behavioural deterioration will occur, and at 20,000ft the normally accepted physiological limit is reached. Only a few minutes of consciousness can be maintained at 25,000ft.

The activities in which people are engaged are significant in two senses. Crew members have operational tasks to carry out, and it is necessary that the standards of their performance should remain at an acceptable level. For this reason, oxygen requirements for crew will be at a higher level than those applicable to passengers. The second aspect of activity level derives from the higher consumption rate of oxygen during the performance of physical activity. Consequently, the period of resistance to oxygen depletion is shorter for active cabin attendants than for sedentary passengers.

Consciousness is significant in terms of the ability to perform useful tasks for the benefit of oneself or others. Since this is not easy to define, the notion of Time of Useful Consciousness (TUC) is used to describe the period during which a person might remain capable of performing a task such as donning an oxygen mask. Several studies have attempted to establish guidelines concerning the length of TUC which might be expected under operational conditions. In one investigation, in which the sudden decompression which had occurred in a McDonnell Douglas DC-10 was simulated in the laboratory, the "cabin" altitude rose from 6,500ft to 34,000ft in a period of 26s and was followed by descent at 5,000ft/min (B27). It was found that healthy young people, representative of the flight attendant population, could be expected to exhibit TUCs of 54s when at rest. However, this period was reduced to 33s when they performed tasks utilising the same energy expenditure as that involved in carrying out cabin duties.

Following exposure to such high altitudes, a period of recovery at or near sea level will be necessary before normal standards of performance can be achieved.

In a pressurised jet aircraft at typical cruising altitude a sudden decompression will necessitate a descent at maximum rate to a safer altitude. As the cabin pressure reaches 14,000ft, oxgyen masks will have been deployed and should be used immediately. Cabin attendants should secure an oxygen supply for themselves without delay, and sit down until the aircraft levels out at which time they should offer assistance to any passengers in need. The aircraft will remain at low altitude, consistent with adequate ground clearance, and land as soon as possible.

It has been observed that passengers are reluctant to use the oxygen masks without receiving explicit instructions to do so. This is likely to jeopardize their wellbeing as time is so critical and flight attendants are not in a position to help them.

A reduction in atmospheric pressure causes the lowering of the boiling point of liquids. At the pressure corresponding to an altitude of 27,000ft the boiling point of water is reduced from the usual value of 100°C to 50°C. A hazard is thereby created when hot liquids suddenly boil producing steam and vapour which could lead to scalding. The greatest danger exists in the galley, but passengers could sustain injuries when, for example, coffee-pots spurt hot liquid.

IN-FLIGHT FIRE

Perhaps the most notorious aircraft accident attributed to in-flight fire took place in the Middle East in 1980. Only seven minutes after take-off, smoke warnings were heard in the cockpit indicating fire in a cargo compartment. Four minutes after this, the Captain decided to return to Riyadh and after a further seventeen minutes, the aircraft landed at the airport and stopped on the taxiway. It was more than three minutes after this that the engines were shut down. No one evacuated from the aircraft; all the occupants died and the aircraft was completely destroyed by fire (P05).

Incidence

The incidence of in-flight fires cannot be specified precisely. Different criteria for reporting incidents and accidents, together with the problems that are always associated with the reporting of incidents which do not involve injury or major damage, have the effect of reducing the reliability of accident statistics in all transport and industrial areas. The problem can be illustrated by comparing the incidence of in-flight fires as reported within one airline over a five-year period (a total of 60) with the incidence as displayed in NTSB data over a ten-year period culminating in the same five years (a total of 40) (C22).

One study showed that, in the period 1960-1976, the total number of reports of in-flight smoke and fire world-wide amounted to 171, or approximately one occurrence every month (M20). Data from the CAMI Cabin Safety Data Bank for the period 1974-1983 showed a total of 206 incidents, nearly two per month, involving smoke and fumes in the aircraft. Nearly two-thirds of these involved an emergency landing (H05). There were 42 incidents of smoke in the cabin reported to the FAA/Incident Data System for the years 1980-1984.

Origins of in-flight fire

While an in-flight fire is not an unusual event, it is nevertheless

Figure 7.1 A training simulation of smoke in the cabin

unusual for such a fire to lead to disastrous consequences. Although
in-flight fire is popularly associated with cabin fire, the evidence
shows that most in-flight fires occur in the engines.   For example,
it was reported that in the period 1964-1983 there were 280 in-
flight fires in air transport aircraft, of which 52 (18%) involved
fatalities.   A total of 2,061 people died.   The location of these

fires is shown in Table 7.3 (R02). It can be seen that while fires in the engine accounted for 71% of the total fires, they were associated with only a quarter of the total deaths. By contrast, fuselage (including cabin) fires accounted for only 17% of the total fires but more than half the deaths.

Considering only the cabin, as distinct from the fuselage, there were nine fatal cabin fires in turbine-powered transport aircraft during this period, in which 776 occupants died and only 47 survived (B23).

Proportionally, this group represents 3% of the fires and 38% of the deaths.

Table 7.3 The origin of fires occurring in flight 1964-83 (RO2)

Location	No. of Fires	No. of Fatal Fires	Occupants Killed
Fuselage (incl. cabin)	48	15	1102
Engines	199	28	538
Wheels	17	2	82
Other airframe	6	1	34
Lightning	3	3	189
Unknown	7	3	116
TOTAL	280	52	2061

Causes of in-flight fires

In the period 1980-1985, 63 incidents of in-flight or ground fire were reported to the FAA Accident/Incident Data System. The major causes of these fires (nearly 67% of the total) were mechanical failure and electrical malfunction. Only four involved smoking activities, nine involved lavatories, and five involved the galley, ovens, or food. In spite of this, passenger behaviour, particularly in relation to smoking in the lavatory, has been widely regarded as a major factor in the causation of in-flight fires. These perceptions are supported by the evidence of reports of incidents of fire, smoke or fumes in the aircraft cabins of one airline over a period of nearly four years (B17). During this period, 61 incidents of active flaming or smouldering combustion were observed in the cabin. As the table shows, the greatest single factor was cigarettes on the seats, followed by cooking equipment and the waste bin in either the lavatory or the galley.

Passengers also bring on board items which are hazardous and

which are listed as "Dangerous Goods".   These include camping gas,
cigarette lighters and aerosol containers.   Boxes of matches and
books of matches have been identified as causes of fire in the cabin
and   the hold.   An American airliner was forced to land   after   the

Table 7.4   Fatal cabin fires 1964-83 (B23)

Fires	Fatalities	Survivors
9	776	47

discovery of a fire under the cabin floor caused by a book of
matches which ignited in a woman's handbag which was on the floor
beside the return air grill. This incident together with other similar
ones influenced the decision of one airline to prohibit book matches
being carried by flight or cabin crew.   In September 1983, a
passenger boarding a McDonnell Douglas DC-10 was found to be
carrying 144 quarter litre bottles of hydrochloric acid and in April
1984, a passenger attempted to carry 10kg of fireworks on board a
Boeing 737.   The large quantities of tax-free spirits generally
distributed throughout the cabin are not classed as "Dangerous
Goods".

Table 7.5   Origin of in-flight cabin fires (B17)

Location	Incidents
Galley equipment (incl. ovens)	17
Cigarettes on seat	20
Waste bin, wc of galley	13
Electrical malfunctions	4
Passenger carry-on bags	2
Matches	2
Malfunctioning oxygen generator	1
Cigarette in lavatory supplies	1
Passenger seat	1

Fire fighting
Typically, in-flight fires are small at the outset and are more
readily controllable than liquid fuel fires.   However, there may be
difficulties associated with the early detection of such fires and
with their accessibility. The major threat to occupants is not from

excessive heat or burning but from toxic gases, which are generated by the combustion of cabin materials, and smoke.   Smoke is likely to increase in intensity as the fire continues to burn, with deleterious effects on the activities of the flight crew and flight attendants in carrying out their tasks.   Smoke reduces visibility, and the toxic gases in the smoke reduce the effectiveness of cognitive functions, causing disorientation and impairing judgement.

The prompt response of cabin attendants and flight crew to the presence of an in-flight cabin fire is critical for survival. The common errors made by crew have included a flight crew member inappropriately leaving the flight-deck with the intention of fighting the fire; delay in proceeding to the nearest suitable airport; shutting down air-conditioning packs; and the deployment of passenger oxygen masks.

When an in-flight fire is reported to the flight-deck, the captain may decide to send a member of flight crew into the cabin to investigate and, possibly, to fight the fire.   This strategy has been criticized on the grounds that it results in reduced manpower on the flight-deck (particularly in the case of a two-man crew) at a time when a full crew complement is highly desirable.   There is also the possibility that the crew member may be prevented by the fire from returning to the flight deck, or may be badly burned in an attempt to fight the fire and thus jeopardize the prospect of a successful landing.

A review of 40 accidents involving in-flight fire indicated that in most of these events, flight crew did not declare an emergency or take any action other than to continue to their planned destination (F09).   In normal operating circumstances, crews are under pressure to reach their planned destination and to do so on time for technical and commercial reasons.   These everyday pressures have, on occasions, been allowed to obscure a positive recognition of the danger inherent in in-flight fire.   The delay in making an emergency descent was cited by the NTSB as a causative factor in an accident in which 23 out of 41 passengers died (N20).

In the same incident, the First Officer shut down the air-conditioning packs in the mistaken belief that these would otherwise aid the spread of fire, though this action was not specified in the smoke removal procedures.   In fact, the resulting reduction in the air circulation in the cabin was considered to have accelerated the accumulation of heat and toxicity of the air.

There has been some discussion about the potential role of passenger oxygen masks in an in-flight fire.   In 1979, a Boeing 727 developed an atmosphere of smoke and fumes as a result of an in-flight bomb explosion.   Flight-deck crew manually deployed oxygen masks for passengers who later "indicated that this action had saved their lives".   A similar incident involving a Douglas DC-9 was recorded in 1980 (S29).   Subsequently however both manufacturers of the aircraft involved issued bulletins indicating that manual deployment of passenger masks without a   decompression would not provide protection as there was no oxygen flow to passenger cabin masks at normal cabin altitudes. It was suggested that the passengers derived some psychological benefit from the use of oxygen masks which reduced the likelihood of hyperventilation resulting from anxiety.

Since oxygen masks supplied from stored gas are ineffective because they deliver oxygen only during a decompression, either a redesign of this system or the use of chemically generated oxygen is required for oxygen masks to operate effectively as life-support systems during an in-flight fire. However, the disadvantage associated with the presence of oxgygen is that it may act as an additional hazard by enriching the air and thus intensifying the fire. This hazard is increased if the distribution lines are burned. Against this must be weighed the argument that the amount of oxygen enrichment would be minimal and thus not hazardous in relation to its life-support capability (B15).

The usefulness of smoke hoods and their role in passenger protection has been topical particularly since an aircraft fire in 1985.   There appears to be a vocal lobby in favour of the mandatory provision of smoke hoods though this has been resisted by the regulatory authorities in several countries on the grounds that other methods of combating the problem are likely to be more effective. An extended discussion of respiratory protection may be found in Chapters Eight and Nine.

IMPACT

In June 1976, a McDonnell Douglas DC-9 crashed at Philadelphia, seriously injuring 86 of the 106 occupants, while attempting to land in adverse weather conditions. The aircraft descended in a nose-up attitude with landing gear retracted, struck the taxiway tail first and slid 2,000ft before coming to rest. During this slide, the tail, together with the engines, separated from the fuselage. Severe damage was inflicted on the bottom of the fuselage, the cabin floor was buckled and 95 seats failed at impact, causing passengers to be thrown about the cabin. The aisle was blocked by the failed seats and by the contents of overhead storage racks. At least 12 passengers were unable to leave the aircraft of their own volition either because of injury or because they were trapped by failed seats (N14).

Survival of an impact depends initially on four main conditions. The first is that the acceleration forces are within the limits that can be tolerated by the human body; the second is that the integrity of the structure in the vicinity of the occupant is maintained; the third is that the restraint system, that is the seats, the anchor points, and the seat belts, remains intact; the fourth is that no lethal injuries result from impacting objects such as the seat in front. Additionally, the resistance of the cabin furnishings and fitments to impact forces, the absence of a post-crash fire, and the proximity of rescue services will aid survival after impact.

Tolerance to acceleration forces
Acceleration involves a change in either the magnitude or the direction of motion.   In common usage, "deceleration" describes a reduction in the magnitude. This is the same as acceleration in the direction opposed to that of the motion. The unit used here to express the magnitude of acceleration is G, the dimensionless ratio of a given acceleration to that produced by the Earth's gravity, the latter having a magnitude of $9.81m/s^2$. This avoids the confusion arising from the use of g, which "has been widely abused in the aerospace medical literature" (S19).

Acceleration alone is sufficient to cause injury or death. Even if the human body were to be so carefully restrained and packaged that no possible damage could be inflicted by the surrounding structures, nevertheless the action of the acceleration forces could, if they were sufficiently powerful, cause death or serious incapacitating injury.

The question then arises of how much acceleration can be tolerated by the human body before death occurs or irreversible injury is inflicted.   It is not easy to answer this question for a number of reasons. Tolerance of acceleration forces is multi-dimensional and the dimensions are interdependent.   The variables involved include the magnitude of the forces, the rate of onset, the duration of the forces, and the direction in which the forces impinge upon the body.   It has been shown that, with torso restraint, humans can tolerate a forward acceleration of 50G at 500G/s rate of onset for a period of 0.25 seconds (S25). Increasing the rate of onset would require a reduction in duration and a diminished magnitude of acceleration for tolerance to be maintained.

The direction in which the forces are applied influences the extent to which these forces are tolerated.   Greater tolerance has been observed to rearward than to forward acceleration. Survival with reversible injuries has been reported for a rearward acceleration of 83G for 0.04s at 3800G/s (U01).

Another reason for the difficulty in specifying impact tolerance is the practical problem involved in obtaining data. Clearly, experiments on human subjects cannot be carried out to determine the limits of tolerance and the evidence from cadavers and animals is not directly transferable to a living human population. The investigation of tolerance below the limits is complicated by the lack of standardization in the definition of tolerance which varies with the investigator from the level where subjective feelings of pain are experienced to the point where various degrees of injury are sustained. It is also important to bear in mind that the data which are available have most often been collected, in common with a great deal of biomechanical data, from fit, young, male (frequently military) subjects.

A third reason for the difficulty in specifying acceleration tolerance is that there are differences in the effects of these forces on different individuals. Tolerance varies between individuals and it varies within the same individual at different times. Characteristics of the individual such as age, sex, physical and mental condition, all affect the ability to tolerate acceleration forces.

A different approach to the determination of tolerance to acceleration forces is by means of an analytic study of aircraft accidents. A review of 77 accidents by the NTSB showed that occupants were surviving crash forces greatly in excess of the US design criteria current at the time (see Table 7.6A). The NTSB summarized the conclusions of a number of relevant organizations that, where the occupant is restrained by a lap belt and where the aircraft and its occupants experience a rate of onset and duration of forces typical of those experienced in survivable crashes (50G/s for 0.1-0.2s), the human body can withstand forces two or three times greater than those to which the designs must conform (see Table 7.6B) (N17).

For a further comparison, it is interesting to note that United States Air Force design recommendations for forward acceleration based on seat belt and upper torso restraint are 45G for 0.1s and 25G for 0.2s (S21).

Table 7.6A The range of forces that the human body can withstand without sustaining irreversible injury when restrained by a lap belt and experiencing a rate of onset and duration of forces typical of those experienced in survivable crashes (N17).

Direction	G
Forward	20 - 25
Downward	15 - 20
Sideward	10 - 15
Upward	20

Table 7.6B   FAA design criteria (1988)

Direction	G
Forward	9.0
Rearward	1.5
Upward	3.5
Downward	6.5
Sideward	1.5

However, the evidence from the study of accidents will not provide tolerance data comparable with those from the laboratory as the acceleration experienced in an aircraft will be complex, involving forwards, backwards, sideways, downwards, and upwards forces as the aircraft impacts with trees, buildings or other features of the terrain before coming to rest. This was recognised in 1988 in the development of standards for seat strength which, in addition to increasing static strength requirements, introduced the requirement for dynamic testing. (See p 130)

Intact fuselage
The value of an intact fuselage lies in the protection it provides against ejection which has a high probability of causing fatal injury. It will also reduce the likelihood of fire entering the cabin and thus allow for longer escape time. "Current fuselage structures are doing a relatively good job of protecting occupants in crashes with large forces" (N17).

Restraint systems
If an aircraft were to impact with high ground at cruise speed, any occupant standing in the aisle would be propelled at the same speed through the cabin with a force that could penetrate the structure of the aircraft. Seated passengers, restrained with a lap belt, would incur very serious, if not fatal, injuries in such a collision. However, at speeds more characteristic of impacts with the ground, the restraint system has some survival value.

The restraint system comprises the seats, the structures which attach the seats to the fuselage, the seat belts and seat harnesses. The orientation of the seats in the aircraft is a factor in survival. The tolerance of individuals in side-facing seats is less than that of individuals in forward-facing seats, which in turn is less than that for individuals in rearward-facing seats. In rearward-facing seats, the forces are better tolerated as they are distributed over a wider area of the body.

Fatalities have resulted where the acceleration forces have been within tolerable limits and the fuselage structure has remained intact but where the seats have broken loose from their anchor points and occupants have been hurled with a force of 20G or more into the seats in front or crushed by those impacting them from the rear.

In an impact, forward-facing lap-belted occupants will continue to move forward until the slack of the belt is taken up.   Then the head and upper torso will pivot forward about the belt, and the arms and legs extend forward.   When the immediate environment contains sharp or solid objects which are then struck by these body parts with considerable force, serious injury will result.   The injuries are likely to be primarily to the head and in addition to the legs and arms.   Other injuries sustained from the restraint system are back injuries, which may result from rigid seat frames or from loose or badly fitting seat belts, and injuries to the chest and pelvis.   Seat belts may also cause internal injury.

The forces with which the occupants' heads and legs impact the seats in front are likely to cause failure in the tie-down structures of that seat row, with the consequence that these seats and their occupants impact with even more force on the seats in front of them.   Thus failure of seats at the rear of the cabin may cause a domino effect of failing seat rows throughout the cabin.

To attempt to attenuate the effects of impact, forward-facing occupants are advised to adopt the "brace" position (bending forward and gripping ankles).   However, passengers whose dimensions are beyond the normal range and those who are pregnant may experience problems in assuming this position.   Particular difficulties may be encountered in some aircraft where, in order to increase passenger capacity, seat pitch may be reduced to less than 760mm (30in).   In these circumstances a variation of the conventional "brace" position must be adopted.

Cabin furnishings
Impact forces have caused internal furnishings and fitments, such as ceiling panels, passenger service units, movie projectors and screens, and magazine racks, to break loose.   Overhead furnishings failed in nearly 80% of the accidents analysed by the NTSB (N17).   When catches on overhead lockers fail, their contents, which may include heavy items such as tax-free bottles or small suitcases, are hurled about the cabin often causing serious injury to anyone in their path.   If these injuries lead to incapacitation or unconsciousness, the chances of survival are minimised in the event of a fire.

Even without the infliction of injury, the broken fitments and the other items may cause difficulties in evacuating the aircraft by blocking the aisles; thereby impeding the route to the exit, and by preventing the exit doors from being opened quickly and easily.

A particular problem arises from the failure of galley equipment which was noted in more than 60% of the sample of survivable impacts examined by the NTSB.   This is because the designated positions of cabin staff on take-off and landing are at exits which are frequently placed adjacent to the galleys.   In this position, cabin staff are vulnerable to the movement of heavy galley units and to the sharp-edged drawers which, when unlatched, can slide open and inflict cuts; burns have been sustained from the splashing of hot liquids.   The contents of galley units may cause injury and impede the evacuation.   Entire galley units have been known to block exit doors so that they could be opened only partially or not at all (N17).

FIRE ON THE GROUND

Aircraft have caught fire on the ground for a number of reasons. Fires may occur as a consequence of overheated brakes, burst tyres, or a runway over-run leading to collision with trees, buildings, airport vehicles or perimeter fences.   An emergency landing away from an airfield may result in fire.   After a survivable impact, the possibility of the outbreak of fire is a major concern. Fires have broken out in aircraft while they were at the ramp during maintenance or cleaning.   Other ramp fires have taken place after engine start-up.   This operation is known to be hazardous and, as a precaution against fire, the presence is required during start-up of an airport employee equipped with a fire extinguisher.

Post-impact fire
In 1965, a Boeing 727, approaching Salt Lake City at a rate of descent three times that recommended by the operator, touched down short of the runway, shearing the landing gear.   After impact, the aircraft skidded for about 27 seconds, covering a distance of about 865m (946yd) before coming to rest.   It was observed that, 3-5 seconds after impact, there was a muffled explosion followed by a bright orange flame which shot out from underneath the aircraft, near the tail in the vicinity of the engines.   By the time the air-craft stopped, it was engulfed in flame to an area forward of the wing.   The fuel line from the wing tank to the engines had been ruptured when the right main gear strut was forced up into the fuselage by the impact; the fuel, under pressure, was ignited and the fire burned through the cabin floor "like a blow torch".   This accident claimed the lives of 41 of the 85 passengers on board (S18).
    It is in the minority of impacts that fire ensues.   The more severe the impact, the greater the probability of fire.   The high number of fatalities in these accidents is not attributable solely to

the effects of fire. For the period 1955-1978, 22% of the passengers involved in all aircraft accidents were killed but only 5% of the total died as a result of post-impact fire (T02).

Post-impact fires can vary from those where massive fuel spillage and ruptured fuselage result in the immediate engulfment of the cabin by fire to those where minimal fuel spillage and minor fuselage damage may allow several minutes for escape. In the former cases, the primary danger is from the intense heat, and death occurs mainly as the result of burning whereas in the smaller fires the lethal gases produced by combustion of plastics are more likely to be the cause of death.

Fire hazards
In a study based on world-wide accident experience in the period 1964-1974, fire hazards were ranked in terms of the likelihood of survival and the number of occurrences (B15). The major fire hazard, which occurred in 50% of world-wide impact-survivable fire accidents, was considered to be fuel release from wing separation. Fuel tank explosions were ranked second as these were a factor in both very severe in-flight and post-impact fires. Moderate fuel spills from damaged tanks and fuel lines causing post-impact fires were ranked third as these were associated with numerous accidents where a high percentage of fatalities was caused by fire rather than impact. Cabin materials, implicated in in-flight fire accidents and contributing to post-crash fires, were ranked fourth. Propulsion and landing system fires and explosions taking place during fuel tank maintenance and refuelling were considered to result mainly in damage with few fatalities.

Experimental fire
Tests carried out to study the behaviour of fire within the passenger cabin showed that it took just over two minutes for a fuel fire started outside an open exit door to culminate in flashover inside the cabin (S06). Flashover is associated with fires in enclosed spaces where materials decompose to produce combustible gases. These gases build up below the ceiling, getting progressively hotter, until they ignite resulting in virtually instantaneous spread of fire throughout the cabin.

No two fires behave in the same way and there is some question about the extent to which the results of such tests may be generalised to fire accidents (A02). It is nevertheless instructive to compare the time available for evacuation in the experimental fire with the impressions of the average airline passenger who believed that there would be about five and a half minutes in which to escape if a fire erupted outside the aircraft after it was on the ground (J10).

Probable fire scenario

The most likely fire accident scenario will progress as follows: wing separation on impact results in the spillage of large quantities of fuel; a mist is formed which is ignited through contact with hot engine parts, electrical sparks or hot electric wires; this ignites the pool of spilled fuel which then spreads to the cabin, eventually entering the cabin interior either through breaks in the fuselage from impact damage, failure of windows, burn-through of fuselage skin, or through opened escape doors; the propagation of the fire within the cabin typically culminates in flashover.

Effects on occupants

The occupants of a crashed aircraft in which fire has broken out may have their attempts at escape handicapped by the effects of heat and smoke. Exposure to high temperatures may cause death from thermal shock and, if more than 50% of the body is involved, from burns. Pain and breathing difficulties caused by intense heat cannot be tolerated for very long. Hot, toxic gases and dense, black smoke are released by the decomposition of materials such as polymerics used for hard and soft furnishings. These gases tend to accumulate near the ceiling and can reduce visibility within a few minutes. They may be hot enough to burn the skin and set clothing alight. Eyes stream and breathing becomes increasingly difficult. The processes of combustion of the plastic materials reduce the amount of oxygen in the cabin from 21%, the normal proportion, to 1% within a few minutes. In the event of flashover taking place, death is the most likely outcome.

Before a door is opened in an emergency, it is important to ensure that there are no signs of fire in the immediate area outside the aircraft. This is for two reasons. Passengers must not be allowed to evacuate into an area which is likely to be dangerous, and any opening in the fuselage will permit the fire to enter the cabin and thus reduce considerably the time available for evacuation, further threatening survival.

Anti-misting kerosene

The importance of the ignition of the fuel mist in the initiation of a post-crash fire has led to attempts to develop fuel additives which prevent misting. Following a protracted programme of research into anti-misting kerosene (AMK) and numerous calls for its compulsory adoption, the FAA and NASA jointly mounted a controlled impact demonstration (CID). This was intended to precede the introduction of the proposed legislation.

In December 1984, a remotely-piloted four-engined transport aircraft was landed in the Californian desert. The CID was intended not only to demonstrate the advantages of AMK in a post-crash fire, but also to provide data upon restraint systems designed to improve impact survivability. Land-based cutters were sited such that the aircraft would be damaged as if in an accident. The aircraft contained a wealth of measuring and recording devices to provide the various data required.

In the event, the aircraft landed a little way short with one wing low and yawed to the left. One of the cutters punctured a wing between the two engines causing a large fuel spillage. A damaged engine provided the continuous source of ignition necessary for AMK to burn. The resulting conflagration brought about a further delay in the compulsory introduction of AMK into jet transport aircraft.

In spite of these circumstances, the CID did provide additional evidence of the advantages of the fuel which burns at a comparatively low temperature, and demonstrated the effectiveness of improved restraint systems.

EMERGENCY GROUND LANDINGS

Under certain conditions, a decision may be made to land an aircraft away from an airfield. Such an event might be brought about by fire on board, fuel shortage, or by a severe Hardware failure. Cabin crew duties will comprise preparing the cabin by stowing all movable equipment, ensuring that exit routes are clear, and securing loose items. In addition, passengers must have their belts secure in upright seats, and selected passengers will be briefed and located to assist with an evacuation. This will take place immediately the aircraft comes to rest, using all available doors.

In previous times, forced landings were a good deal more frequent and could be occasioned by adverse weather affecting aircraft flying at relatively low altitudes. Over well-populated areas, the most useful aid might have been a railway timetable. On "long-haul" routes, aircraft could spend long stretches of time in isolation from contact with agencies on the ground or with other aircraft. Consequently, an unscheduled landing might have led to the necessity for crew and passengers to survive for a considerable period of time prior to the arrival of assistance.

Numerous developments in air transport have brought about some alleviation of the problem. The mechanical reliability of aircraft and the margins of safety required of their performance have reduced the frequency of unscheduled landings. The higher cruising altitudes not only reduce problems associated with weather but, combined with increased speeds, place aircraft in easier reach of an airport in the event of an emergency. The enhancement of radio systems has improved both communication and navigation, thus reducing the necessity to search large areas of terrain following a forced landing. Search and rescue facilities are effectively co-ordinated on an international basis. Nevertheless, there remains some necessity to prepare for survival following the landing.

Hostile environments
Four broad categories of geographical environment have traditionally
been identified as those which require specific preparatory
equipment and procedures to facilitate survival. These are arctic,
jungle, desert, and ocean. Two of these which affect the majority
of long-haul carriers, namely arctic and ocean, still require on-board
preparatory resources in view of the urgency of the initial
protection necessary for survival.

Arctic survival
In addition to the problems associated with surviving a crash, and
the possibility of post-crash fire, arctic survival requires prompt
attention to protecting personnel from the extreme cold. In the
event that the fuselage survives, it obviously provides some
protection from the cold. Aircraft will carry some amount of
supplementary arctic clothing, together with such items as shovels
and stoves. In addition, the cabin furnishings can be used to
improvise additional items for warmth. Cabin attendants should be
vigilant to ensure that everyone is kept as warm and dry as possible
until help arrives.

DITCHING

Ditching, in the sense of a forced, controlled "landing" on to water,
is infrequent and most likely to be due to an insufficiency of fuel
to remain airborne. Because it implies a decision on the part of
the pilot and thus some preparation time, however short, pre-
meditated ditching is to be differentiated from an unforeseen
emergency resulting in an impact which takes place on water rather
than on land. The latter typically takes place close to the point
of departure or landing.

Incidence
Unplanned landings into water, although infrequent, occur more
often than premeditated ditchings. In the period 1976-1987, there
were on average three impacts in water each year world-wide,
ranging from seven in 1978 to one in 1983 (E04).
   In January 1982, a Boeing 737 with 74 passengers and five crew
took off from Washington National Airport in conditions of snow
and ice. Moments later, mainly because of icing on the wings, it
crashed into a bridge and plunged into the River Potomac. Only
four passengers and one crew member survived (N18). By contrast,
in May 1970, after diversions and changes of plan on account of bad
weather, a McDonnell Douglas DC-9 ran out of fuel over the

Caribbean.   The pilot announced the possibility of ditching some ten minutes before impact with the water, during which time preparations were made in the cabin.   Three extremely violent impacts were felt, throwing passengers, both belted and unbelted, from their seats and in some cases to their deaths.   The aircraft floated for about five minutes and the majority of survivors escaped through an over-wing exit. The survivors grouped on and around an evacuation slide which had been inflated by the navigator until rescue helicopters arrived to winch them to safety (N04).

Factors affecting survivability

If an aircraft disintegrates on impact, a ditching is unlikely to be survivable.   If an aircraft remains intact, then survival of the impact is likely, though subsequent death may result from drowning or exposure.

Survivability in those cases where the aircraft does not disintegrate depends on a number of factors.   The body-build and physical condition of the passengers, the temperature of the water and the length of time before rescue are all important in relation to survival.   The availability and correct use of survival equipment and the design of the aircraft also affect the outcome.

The passengers most vulnerable to the effects of ditching and to the potentially hostile environment are the very old, the very young, the previously disabled and those disabled by injury incurred as a consequence of ditching.   These groups will find evacuation of the aircraft more difficult and are less fitted for coping with the effects of submersion in water.

The temperature of the water is likely to be the most critical of the environmental factors.   Cold water is more chilling than cold air as heat passes more quickly from the body to the water.   The danger is death from hypothermia.   The core temperature of the body must be maintained at, or very close to, 37°C, and deviations in either direction may be fatal depending on their duration and amplitude.   When core temperature drops below 35°C, body temperature regulation is likely to go out of control, heart beat is irregular, breathing is slow;   there is constriction of blood vessels to the skin and to the extremities.   This will result in a loss of the use of hands and forearms and consequently passengers' grip on flotation cushions, pieces of wreckage, or other people, will fail.   Loss of consciousness is preceded by irrationality which is inimical to self-preservation. At core temperatures of 28°C survival must be in doubt and at 25°C, death is imminent.

There is a relationship between time of exposure and water temperature which is further complicated by body mass, amount of insulation from clothing, and the extent to which the body is submerged in the water.   Very fat people will survive longer in low temperatures than those who are more normally proportioned.

Protective clothing will extend survival time. For example, a wet-suited diver could survive for many hours in cold water. Without protection, however, in water close to freezing, survival will not be prolonged beyond an hour and may be considerably less. The colder the water, the greater the rate at which the core temperature drops. While the hazards of hypothermia from cold water may be absent from more temperate climes, there may be instead the hazard of sharks in the water.

It is clear therefore that the chances of survival will be increased if the period of submersion in the water is minimised as far as possible. This may not be easy. After a ditching in the Caribbean, ninety minutes elapsed before the first passenger was winched to safety (N04). Even if rescue is swift to arrive, the time taken to winch perhaps more than one hundred individuals to safety may be longer than the survival time of many passengers. In the ditching referred to above, a further ninety minutes were required before all 41 passengers were winched to safety.

A wide-bodied aircraft is likely to have more buoyancy than a narrow-bodied one, thus allowing more time for evacuation. However, more passengers are carried in these aircraft. A low-wing is more advantageous in ditching than a high-wing configuration. The former allows the aircraft to fly at lower speeds near the surface of the water. This permits a softer landing, and therefore less consequential damage to the fuselage. Because the fuselage rides high in the water, supported by the wings, passengers are able to utilise the exits and to launch the life rafts from the wings. The high-wing configuration, on the other hand, particularly when accompanied by a high tail, is more likely to sink rapidly on impact to wing level such that the fuselage is immersed.

Survival equipment
The design of survival equipment, its accessibility and availability when required, and the comprehensibility of the instructions for its use clearly have a bearing on the potential for survival. The equipment available for survival in water consists of flotation cushions, life jackets, life rafts, and some inflatable slides which, when disconnected, may be used as life rafts.

TERRORISM

Civil aviation, in common with most other civilian enterprises, is vulnerable to the hazards presented by terrorists operating outside the rules which govern civilized society.

The earliest spate of attacks upon civil aircraft occurred during the 1950s when aircraft were hijacked to convey refugees from eastern Europe to the West. During the late 1950s, hijacking was employed by those wishing to escape from Cuba to the United States, and after the fall of Batista, hijacked traffic flew in the reverse direction into Cuba. Although in the majority of cases the objective was to achieve no more than one-way transport, considerable danger surrounded the operations and several accidents resulted.

The subsequent escalation in attacks on aircraft was associated with a change in motive, involving the taking of hostages for political goals. Most incidents have been connected with the affairs of states in the Middle East.

In addition to hijacking, civil aircraft have been attacked by ground-based missiles, and by fighter aircraft, and have fallen victim to explosives being detonated on board. Whilst most of the assaults have been organized by political pressure groups, occasional attacks have been made by individuals, sometimes of bizarre disposition.

Defence against terrorist attacks lies mostly outside the scope of personnel in the aircraft cabin. Some basic security duties of contracting states are defined in the ICAO Annex 17. Many states are signatories of ICAO security conventions held in 1963, 1970 and 1971. Numerous other international organizations have promulgated policies concerning terrorism.

Airport security procedures play the major role in ensuring that weapons and explosives are not taken aboard aircraft. In view of the large number of personnel having access to aircraft, in addition to passengers boarding for a flight, the security task is formidable. Standards are variable.

In the event of an armed attack, flight crews are severely limited in their options. Some training is given, the details of which are, of course, confidential. At present, no attempt is made to provide a briefing for passengers. Experience has shown that compliance with the demands of hijackers is the best form of defence. Passengers should attempt to remain calm, avoid making themselves conspicuous in any way, and be prepared to tolerate the various hardships imposed upon them.

8   Hardware in Emergencies

> All well-designed Hardware will have safety incorporated as a
> fundamental design criterion rather than having certain
> features appended as cosmetic extras.   In considering
> emergency situations, attention must be paid both to items in
> general use within the cabin, such as seats and furnishings,
> and also to those items, normally described as "safety
> equipment", which are available for use during an emergency.

In Chapter Three, an account was given of some ergonomics
features of the cabin pertaining to the accommodation of passengers
and the requirements of the cabin crew during normal operations.
Consideration of the emergency situations which can arise produces
new sets of design constraints which must be met in order to
achieve acceptable levels of safety.   Clearly, many compromises
must be made.   Whilst safety is universally regarded as a high-
priority item, practical solutions must also take into account the
necessity of achieving commercial viability.

SEATS

Seat pitch
Seat pitch (the distance between similar points on seats in the
adjacent row forward or behind, measured horizontally) varies in
different aircraft.   It also varies according to class of passenger
accommodation.   Some First Class sleeper seats may have a pitch
in excess of 1570mm (about 5ft) whereas the pitch of low-cost
seating may be less than 710mm (28in).
   Seat pitch has obvious implications for the comfort of the seated
passenger and for ease of getting into and out of the seat.   It also
has implications for safety, particularly in the case of obese or
handicapped passengers who are likely to experience difficulties if
they occupy seats other than those adjacent to an aisle.   A seat
pitch of 34in (864mm) has been described as "restrictive" in this
connection (B13).   In a simulation study, when handicapped subjects

were seated by a window in a row of three seats with a 34in pitch, they took up to 50% more time to reach the exit than when they were seated beside the aisle.

Seat pitch also affects the risk of injury in the event of impact. In forward-facing seats, where restraint is provided by a lap belt, which may itself inflict serious abdominal injury if it is incorrectly adjusted, occupants are likely to be killed in a crash of 5G or more as a result of striking their heads on the seat in front. The padding on the seat-back is insufficient to absorb the energy and an impact of even less force may result in head injury or concussion, reducing the ability to escape. Fold-up tables on the seat-backs exacerbate the problem. Clearly, the design of future seats incorporating video screens and stereophonic sound systems has serious implications for safety.

One solution to the problem of potential head injury is the use of shoulder harnesses which are likely to reduce both head and facial injury even with a seat pitch of 810mm (see Figure 8.1). However, these are unlikely to be popular with the travelling public. Furthermore, there remains the potential problem of injuries sustained by hands and arms, and by feet and legs which result from flailing limbs making forceful contact with other seats in the vicinity. Injuries of this sort may, in turn, cause difficulties for passengers trying to release seat belts with injured hands, or attempting to evacuate the aircraft with injured legs. To avoid these injuries, the seat pitch would have to be uneconomically large.

The special problem of seat pitch in relation to access to emergency exits is discussed later.

Seat location
The location of seats in relation to exits and the implications for survival of the seat-to-exit distances are discussed in the section on Exits. Seats in the rear of the aircraft may be safer in terms of post-crash survival than those in the front. An early study showed that passengers seated in the rear of the aircraft were seven times more likely to survive, three times more likely to have no serious injury and three times more likely to have no injury than those seated in front of the leading edge of the wing (W05). More recently, a review of more than 90 accidents was conducted to determine the location of the safest seats. In 21 cases, seat location and injury appeared to be related. These included 14 which occurred during approach and landing, four which occurred during take-off, and three water landings. In two-thirds of these accidents, those seated at the rear of the aircraft incurred fewer injuries than those seated at the front (J13). In part, this is because the front of the aircraft is likely to impact first, with the greatest force, but it is also due to the widespread use of fuel tanks in the wings and consequently the risk of fire in that area.

Where the aircraft tail impacts the ground first, more casualties occur, as may be expected, at the rear.

Seat construction and tie-down

By the early 1930s, it was apparent that regulations were needed demanding that seats should no longer be placed loosely on the floor of the aircraft or anchored in only a casual way. The US Department of Commerce required, in 1932, that "seats and chairs should be firmly secured in place". Later Civil Air Regulations (CAR) defined this security by requiring that seat fixings must withstand the same load factors as those required of the aircraft structure, whilst the seat was occupied by a person weighing up to 170lbs (77kg) (F08).

These aircraft load factors, dating back to 1926, have only slowly been modified over the intervening years, as shown in Table 8.1. Obviously, such an approach to the problem of seat anchorage owes nothing to the data concerning human tolerance to acceleration (see Chapter Seven).

The regulations required that safety belts sustain the emergency landing loads and that seats sustain the most critical of the emergency landing loads, flight loads, landing loads and pilot-induced forces. In 1952, a safety factor of 1.33 was applied to each load factor for seat and safety belt attachments and the forward emergency landing load factor was increased from 6G to 9G (Table 8.1B).

A report by the US National Transportation Safety Board (NTSB) dated 1981, which examined cabin safety, recorded that in 1962, the Bureau of Aviation Safety of the Civil Aeronautics Board (CAB, predecessor of the NTSB) analyzed an accident in which 28 passengers died and concluded that seat failures probably prevented at least some of these individuals from evacuating the aircraft (N17). The CAB went on to recommend to the Federal Aviation Administration (FAA) that "studies relative to crash load factors and dynamic seat testing criteria ... be expedited toward the end of improved safety in this area at the earliest date".

A Notice of Proposed Rulemaking was issued by the FAA in 1969 to increase the upward and sideward load factors of passenger transport aircraft to 4.5G and 3.0G respectively. A Notice in 1975 proposed the addition of a rearward load factor of 1.5G for transport aircraft. However, these Notices were withdrawn after comments that the proposed changes were unrealistic and unnecessary, and that the resulting increase in weight resulting from the stronger anchoring components would be an undue economic burden.

The NTSB report noted that "In 1962, regulations regarding crash forces were already about ten years old; today, 30 years after the standard was established and almost 20 years after the first

recommendation, these regulations have yet to be changed, and
seats and other cabin furnishings continue to fail in aircraft
accidents regardless of the severity of the impact" (N17). The

Table 8.1   The load factors to be accommodated by seats in
transport aircraft

A   1946 Part 04b Civil Air Regulations

Direction	G
Forward	6.0
Downward	4.5
Upward	2.0
Sideward	1.5

B   1952 Part 4b Civil Air Regulations

Direction	G
Forward	9.0
Downward	4.5
Upward	2.0
Sideward	1.5

C   1981 NTSB recommendations

Direction	G
Forward	20 - 25
Downward	15 - 20
Upward	20
Sideward	10 - 15

Table   8.1 (cont.)

D   1988 (FAR Part 25)

Direction	G
Forward	9.0
Downward	6.5
Upward	3.5
Sideward	4.5
Rearward	1.5

regulations required that the occupant "be given every reasonable chance of escaping serious injury in a minor crash landing" when experiencing the forces shown in Table 8.1B.

The NTSB further reported that investigations into the tolerance of lap-belted occupants experiencing a rate of onset and duration of forces typical of those experienced in survivable crashes showed that these occupants could tolerate forces very much greater than those provided for in the regulations (see Table 8.1C).

The NTSB report reviewed 77 survivable or partially survivable accidents to passenger-carrying aircraft which had taken place over the ten-year period 1970-1980. In more than 80% of the accidents, the seat or restraint systems failed. These failures were found to occur typically in the seat legs and the seat-to-track attachment points. Often, deformation of the cabin floor resulted in seat tracks breaking which in turn led to other failures of the seat attachment points and seat legs. Features of the seat such as the seat back, seat pans, frames, arm-rests and tray-tables also failed. In 22% of the accidents, some component of the restraint system failed, mainly in the belt attachment rather than the webbing belt material.

In a crash, a passenger in a forward-facing seat equipped with a snugly fitting lap belt will "jack-knife" about the pelvis (see Figure 8.1). The head and arms will be thrown forward and the legs will swing forward to strike the seat in front typically about 75 - 150mm (3-6in) above the ankle joint. The impact forces exerted by these body parts are sufficient to cause serious, incapacitating injury.

When the impact results in seat failure, the occupants are thrown, with a force of possibly 15 or 20G forward, into adjacent seats, on to the floor, pinned between seats, between the floor and seats, and between seats and the side wall. Serious, sometimes fatal, injuries involving skull fractures, spinal fractures, brain damage, crushed chests, ankle and leg fractures, are likely to result. Even in the fortunate event of avoiding serious injury, there is a strong likelihood of passengers being trapped by the wrecked seats and thus unable to make an escape.   A further hazard is the lap belt which can cause severe internal injury when it is not fitted as tightly as possible.

The call for increased load factors led the FAA to embark upon a research and development programme involving computer modelling, controlled impact demonstration, and accident data analysis (F08). The results of this programme were considered by the FAA to have demonstrated the adequacy of current regulations for occupant protection.   Where seat performance was inadequate, this was found usually to be related to cabin floor displacement and excessive lateral inertial loads.   These deficiencies were considered to be eliminable primarily by the establishment   of dynamic test standards

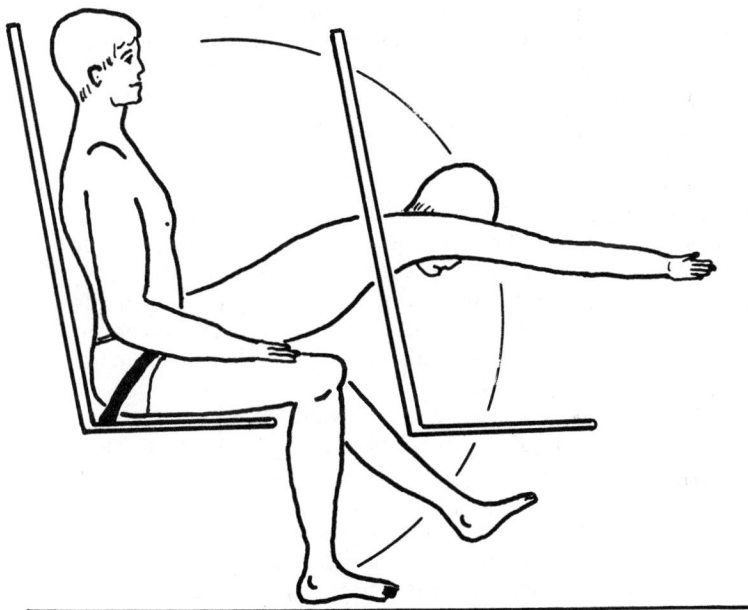

Figure 8.1 The effect of impact upon a passenger with a tightly-fitting lap belt. The arc indicates the boundaries within which all surfaces are required to be smooth and padded.

providing the same level of impact injury protection and structural performance as that provided by the   aircraft itself.

As a consequence of this programme, the FAA introduced

amended seat strength regulations incorporating dynamic test standards (FAR 25.562). Two dynamic test conditions relating to a floor-level pulse were selected; one simulated ground impact following high-rate vertical descent, the other simulated horizontal impact with ground-level obstruction. The former emphasises occupant vertical loading and is concerned with spinal injury; the latter provides an assessment of the occupant restraint system and the structural performance of the seat.

The new seat standards, the first since 1952, came into effect in June 1988 for newly certified transport aircraft. Retrofitting of existing aircraft must be carried out within seven years. The dynamic test standards require that for a combined vertical and forward velocity, peak floor deceleration must occur not more than 0.08s after impact and must reach a minimum of 14G. For a combined forward and lateral velocity, peak floor deceleration must occur not more than 0.09s after impact and must reach a minimum of 16G. Within these limits the seat attachments must remain effective and there must be no deformation sufficient to impede rapid evacuation.

The dynamic tests are to be conducted using a 170lb (77kg) dummy. The weight of 170lb, representing the "average" adult was chosen because designs based upon this value provide the maximum injury protection for the widest range of occupant weights. Tests with this dummy must demonstrate that the lap belt remains on the pelvis and that the maximum compression between the pelvis and the lumbar vertebrae is not greater than 1500lb (680kg).

Further requirements relating to obstructions in front of the seated passenger define the maximum permissible compressive load upon the femurs at 2250lb (1021kg) each and specify the protection necessary against head injury.

For the first time, a rearward design requirement was introduced and the sideward requirement was increased by a factor of three (FAR 25.561) (Table 8.1D).

The NTSB suggested on the basis of data on human tolerance to acceleration that seats be required to withstand static loads greatly in excess of those incorporated in the 1988 rules. However, the FAA argued in response to this that consideration of pulse duration and the changes in velocity appropriate to 20G would result in increased impact velocity and less crash energy. Because energy is a critical factor in demonstrating both the structural performance of the seat and restraint system as well as crash energy protection, such criteria would not provide a higher level of safety (F08). In addition, the floor would have to be strengthened to retain the seats. Clearly, there remain issues still to be resolved.

Seat orientation
Rearward-facing seats provide a safer alternative to the

conventional configuration as greater acceleration forces can be sustained in such seats than in either forward- or sideward-facing seats. The military favour rearward-facing seats, designed in the case of the United States Air Force to withstand 45G.   An American Air Force study of survivable air transport accidents showed that the incidence of injuries to passengers facing forward was seven times greater than for passengers facing to the rear (M25).   In a ditching accident, three unrestrained, rearward-facing flight attendants escaped injury while forward-facing, lap-belted passengers sustained injury (N04).

It is not feasible, however, simply to change the orientation of currently available seats.   The greater stress placed on the seat back resulting from the distribution of the loading over the entire back area of the seated occupant demands correspondingly greater strength of seat construction; the seat attachments must also be considerably stronger;   and there is a need for the addition of an energy-absorbing head-rest.   Rearward-facing seats are thus likely to be much more expensive to construct than forward-facing seats, though the point has been made that advancing technology has already permitted the construction of a 16G seat at a lower cost than the 6G seat it replaced (S21).   However, there seems little likelihood except in the event of regulation that such seats will be introduced into civil air transport.

The apparent willingness of passengers to try out rearward-facing seats expressed in answers to questionnaires is not, of course, firmly based upon experience of such seats in which the "seat-belt-hanging effect" on take-off and climb is the inevitable and uncomfortable consequence, and which is unlikely to be popular (B14).   This could be avoided only by reducing angles of rotation and climb, which would be operationally unacceptable.   However, if rearward-facing seats were to be required by regulation, it seems probable on the basis of past studies that such seats would become acceptable within a short time (S21).

Seat upholstery

Seat upholstery has been identified as a major factor in the causation of fatalities occurring in "survivable" accidents.   This is because of the toxic fumes which are given off when polymeric materials used in seat cushions are ignited.   These materials are also used in curtains and carpets (soft furnishings) and in interior walls, bulkheads, and ceilings (hard furnishings).   The use of fire-blocking materials in aircraft seats has long been resisted because of the associated weight penalties, which in the case of a Boeing 747 could amount to half a ton.   However, new regulations came into force in the US and the UK in July 1987, requiring that the foam cushions be encased in fire-blocking materials under the decorative outer cover. (For a fuller discussion of toxicity, see p.158).

Crew seats

Jump seats are provided for flight attendants to sit upon when the aircraft is taking off or landing. These seats must be situated near to the floor-level emergency exits. They should also be located to provide a direct view of the cabin area for which each flight attendant is individually responsible, though without compromising the requirement to be near to the emergency exits, and the probability must be minimized of the occupants suffering injury or being struck by items dislodged in the galley or from a stowage compartment or serving cart. In a fatal accident which took place during take-off but while the aircraft was still on the ground, two of the four flight attendants were seated in a position from which they were unable to see the section of the cabin for which they were responsible because of the intrusion of a bulk-head (A02).

Jump seats may be either forward- or rearward-facing but not sideward-facing, and they must be equipped with an energy-absorbing rest designed to support the arms, shoulders, head and spine. A lap belt and shoulder harness with a single point release are also required which can be secured when not in use to prevent interference with rapid egress in an emergency. Jump seats are designed to fold back when not in use so that they do not interfere with the free use of aisles and exits. When an airport collision required the immediate emergency evacuation of a Douglas DC-9, the failure of a jump seat to retract to its stowed position created an obstacle to those passengers attempting to exit through the nearby door (N07).

Since trained flight attendants play a critical part in successful evacuation, it is clearly important that they are uninjured in a survivable accident. Unfortunately, flight attendants have been injured in crash landings as a consequence of the inadequacies of the design and sometimes the position of the jump seat. Positions close to the galley (see below) are best avoided. It is the view of one flight attendant that the incidence of injury to flight attendants could be avoided if jump seats were provided with proper belts, harnesses and head paddings and that the seats themselves were anchored securely on the floor rather than attached only to the bulk-head (D11).

Shoulder harnesses

Tolerances of up to 50G have been observed in young males in forward-facing positions using shoulder harnesses (S25). Shoulder harnesses also provide greater protection in the event of impact. The majority of aircraft crash injuries (70%-80%) involve the face and head as a consequence of the upper body pivoting around the lap belt. The introduction of a shoulder harness designed integrally

Figure 8.2 A folding crew seat mounted on a bulk-head adjacent to an exit. Above the seat is the container for drop-out oxygen masks, and below is storage space for items of safety equipment.

with the lap belt would clearly reduce a large proportion of injury. In a ditching accident, the three flight-deck crew who were maximally restrained with seat belts and locked shoulder harnesses escaped injury, while the majority of the passengers sustained injury (N04).

FAA regulations governing the protection of occupants from head injury require the provision of a safety belt and at least one of the following conditions:

. provision of a shoulder harness
. the elimination of injurious objects within the strike
  envelope of the head
. the provision of an energy absorbing rest that is
  capable of supporting the head and upper torso.

The NTSB report of 1981 asserted that the requirement to protect occupants from head injury is not being met (N17). With forward-facing seats, forces are applied mainly from the front and thus it is not appropriate to protect the occupant with an energy-absorbing rest to support the head and upper torso.   Shoulder harnesses are currently unavailable for civilian passengers. Injurious objects are not eliminated from the strike envelope of the head when protection is provided only by a lap belt. Even seat backs padded with energy-absorbing material can inflict injuries on the head and face. Thus, although forces of 30G are survivable in forward-facing seats, much lower G forces may nevertheless prove fatal because of head injury resulting from impact with the seat in front.   While this risk could be considerably reduced with the adoption of a shoulder harness in addition to the lap belt, the implications of the installation of a shoulder harness for passengers in terms of cost and of acceptability are, however, considerable. There remains, too, the risk of neck injury when the head is unrestrained.

Jump seats which often face to the rear of the aircraft are equipped with shoulder harnesses. The importance of the correct fitting of these has sometimes been overlooked (C04). A shoulder harness is only effective while it remains in contact with the occupant's shoulders, thus restraining the occupant from forward movement.   This in turn depends upon the geometry of the shoulder harness and attachment points. An investigation into the restraint system for cabin attendants showed that the maximum shoulder harness angle was 35° but that at least one user installed the harness at an angle approaching 80° (C04). The consequence of this is that flight attendants will slide out of the restraint system in the event of impact.

Seat belt release
It is clearly important that seat belts be released quickly when the signal is given to evacuate the aircraft in an emergency.   There are problems, however, with the design of the release which, because it is on a seat belt, is expected to operate in the same way as a car seat belt release. Even frequent air-travellers have wasted valuable evacuation time in attempting to open their seat belts "by pushing a nonexistent button on the face of the buckle" (M06).

Child restraint
Child restraint is problematical in relation to impact. The seat belts provided in the aircraft seat are not suitable for protecting children. The centre of gravity of an adult is located close to the belt but that of a small child is about 102-127mm (4-5in) above this, leading to the likelihood of the child being rotated out of the seat over the seat belt during deceleration. The introduction of

criteria for approved child restraint systems in the United States in 1982 goes some way to solve the problem for children whose dimensions (40ins and 40lbs; 1m and 18kg) fit that of the child restraint device but leaves unsolved the problem of both larger and smaller children.   In the United Kingdom, young children under two years of age must be accompanied by an adult on whose lap they sit and to whom they are secured by an extension seat belt. Children older than two years in their own seats are secured by the standard seat belts.

EXITS

In the event of an emergency such as a fire or an emergency landing, rapid escape from the aircraft is of critical importance. For this reason the statutory bodies of each certifying state pay great attention to the design and use of emergency exits.   Aspects covered include the number of exits to be provided, their size and shape, the location of exits, the access provided to them, methods of opening and closing the doors, and the escape facilities available to evacuees.

Types of exits
Five standard types of exit have been defined. The dimensions of these are set out in Table 8.2 and illustrated in Figure 8.3.   Of these, Type A and Type I (floor exits) will serve in ordinary use as passenger doors.   In addition to these five types, further exits may be available either through the tail cone of the aircraft, or in the form of ventral exits through the pressure shell and the bottom rear fuselage skin.   These additional types will be acceptable for emergency use provided that certain requirements can be met.

Number of exits
"There shall be sufficient suitable exits to facilitate the rapid escape of all occupants in the event of an emergency alighting" (BCAR D4-3 4.2).   The actual number necessary to meet this requirement will depend upon the number of occupants and the type of exits provided.   Some typical data are shown in Table 8.3.   In the case of exits of an unusual type, such as those installed in the tail cone of an aircraft, the authority may require a demonstration that certain acceptable levels of egress may be achieved.

Location of exits
Exits must be available on each side of the fuselage. They will be distributed in a way to satisfy the certification authority. Some passengers will always have to go further than others to reach an exit.   In the aftermath of a controversy which arose following   the

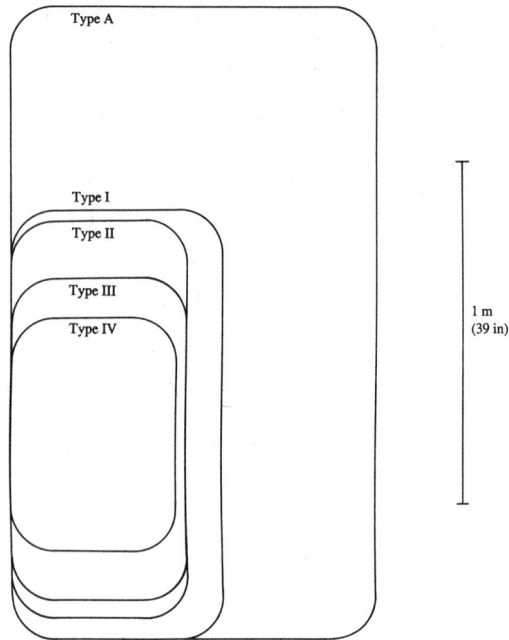

Figure 8.3 Relative minimum dimensions of the five types of emergency exits.

Table 8.2 The minimum size of passenger emergency exits and the maximum step up to the sill.

Type	Min. height		Min. width		Max. step	
	ins	(mm)	ins	(mm)	ins	(mm)
A	72	(1830)	42	(1067)	0	(0)
I	48	(1220)	24	(610)	0	(0)
II	44	(1118)	20	(508)	10	(254)
III	36	(914)	20	(508)	20	(508)
IV	26	(660)	19	(483)	29	(737)

approval of the blocking of the over-wing exits in the Boeing 747, the FAA proposed that the maximum distance of any seat row to the nearest exit should be 30ft (9m) and the maximum distance between exits 60ft (18m). Such a regulation makes no concessions to the variable density seating which is to be found in most aircraft. Whilst 60ft might accommodate about 80 First Class passengers, the same dimension might provide for about 110 Business Class or about 220 Tourist Class passengers in typical configurations.

Access to exits
"Easy means of access to the exits shall be provided to facilitate use at all times, including in darkness; exceptional agility shall not be required...." (BCAR D4-3 4.2.5).   For aircraft having a capacity of 20 or more passengers, the main aisle must be at least 15in (381 mm) wide at floor level increasing to 20in (508mm) above a height of 25in (635mm).   Additional space, in close proximity to the main exits, must be provided to allow a member of the cabin crew to assist passengers out of the aircraft and on to the escape slide.

Access to escape hatches situated beyond a row of seats presents more problems.   Recent revisions of legislation in the UK, resulting from experience of egress in emergencies, have specified dimensions to ensure improved access to Type III and   Type IV emergency exits

Table 8.3   Emergency exits in relation to seating capacity

Passenger seating capacity	Emergency exits each side of the fuselage			
	Type I	Type II	Type III	Type IV
1 - 9	-	-	-	1
10 - 19	-	-	1	-
20 - 39	-	1	1	-
40 - 79	1	-	1	-
80 - 109	1	-	2	-
110 - 139	2	-	1	-
140 - 179	2	-	2	-

Additional emergency exits (each side of fuselage)	Increase in passenger seating capacity allowed
Type A	100
Type I	45
Type II	40
Type III	35

(C16). The new ruling is illustrated in Figure 8.4B. In order to discourage approaches to these exits over seat backs, and to maintain the minimum access width, the seats forward and aft of the exit must not recline and their resistance to breaking forward under a load must be considerably strengthened.

Type III exits require a step up of not more than 20in (508mm) and a step down on to the wing of not more than 27in (685mm). Elderly and disabled passengers are likely to have considerable difficulty with dimensions approaching the maximum (B21).

Obviously it is important that all potential exit routes are kept clear of obstruction. In addition, care must be taken in the design and operation of an aircraft to minimise the risk of obstructions being created by objects breaking loose as the result of impact or turbulence. Service carts and passengers' luggage serve as examples.

Operation of emergency exits

"The means of opening emergency exits shall be rapid and obvious and shall not require exceptional effort" (BCAR D4-3 4.3.6). The types of exit most commonly encountered on passenger aircraft are Type A, Type I, and Type III. Type A doors, as found on widebodied aircraft, are motor-driven in normal use and may be opened by turning or lifting handles which are located on the door or beside the door.

Type I doors are usually opened by rotating a handle, mounted centrally on the door, through 180° in the direction indicated by a red arrow. The direction of rotation will normally depend upon which side of the aircraft the door is located.

Type III exits, which are normally window hatches located over a wing, are more likely than other types to be opened by a passenger because flight attendants are more often stationed elsewhere beside the doors. The hatch, which may weigh about 22kg (50lb), is designed to be removed by an individual in either a standing or a sitting position by pulling a handle at the top. Use of this handle together with one at the bottom allows the hatch to be lifted clear of the exit and then pushed outside the aircraft. However, the weight of the hatch and the awkwardness of the action of pushing it out of the aperture do not make this an easy task for any but a young, strong male. The difficulties in removing the hatch may, however, be somewhat reduced as a consequence of the mandatory increase in space at the Type III exit (see page 146).

If, after a door has been opened, it is apparent that the exit is unusable because of fire, safety is enhanced by re-closing the door (see Exits and external fire, p. 149)

Problems in the operation of exits
Reports from flight attendants suggest that there are problems for
passengers in operating the mechanisms which open the doors and
windows, although the regulations state that all exits must be
capable of being opened on either side by unskilled personnel (M28).
Problems would appear to be inevitable where there is no
standardized method of opening these exits and where, even in the
same aircraft, some handles rotate clockwise and others anti-
clockwise. Standardization of the opening mechanism would simplify
crew training and passenger briefings by avoiding confusion,
particularly since placarded instructions are unreadable in the dark
or in smoke.   The lack of standardization in the opening
mechanisms of exits is likely to result in lost time in an
emergency.
   There are major differences in the maximal forces applicable by
males and females. In one study, female subjects were shown to be
able to pull with 40-80lbs (178-356N) of force and males with 100-
160lbs (445-713N) of force on emergency operating mechanisms
(M08).   By accelerating the body and jerking the handle, male
subjects were capable of applying 300-340lbs (1334-1512N) of force.
Twice the amount of rotational force could be applied by men than
by women.   The workforce in the cabin is predominantly female
and thus the amount of force required to operate doors must be
considerably less than would be the case if the doors were to be
operated solely by men.   Excessive force requiring the effort of
several male personnel is likely to be necessary if the exit
malfunctions for any reason.   Malfunction may be due to damage to
the fuselage on impact, or from poor quality maintenance.

Information display
Regulations require that all emergency exits be clearly marked (eg
FAR 25.811).   In addition, instructions for their opening must be
provided on, or close to, the exit.   Where rotary movement of a
handle is necessary, the direction of rotation is required to be
indicated by a large red arrow having the word "Open" near its
head. The operating handles of Type III and Type IV exits are
required to be self-illuminated.

From exit to ground
Regulations require that all floor level exits more than 6ft (2m)
from the ground are provided with equipment (normally slides,
sometimes stairs) to assist the passengers to descend.   Egress from
Type III exits (windows) requires passengers to lower themselves
from the wing to the ground, a distance typically in excess of 2m.
Because of the increased risk of injury involved, these exits are not
intended to be utilized except during an emergency.   Some aircraft,
such as the McDonnell Douglas DC-10, have slides attached to the

overwing exits.

Requirements and experimental studies

As part of the process of aircraft certification, it must be demonstrated to the authority that an evacuation of passengers and cabin crew can be achieved within 90 seconds. The constraints imposed upon such a demonstration are defined in BCAR D4-3 4.5.2 and in FAR 25.803. These include the following:

1. Use can be made only of the minimum required number of exits on one side of the aircraft
2. Each exit and internal partition must be configured as for a normal take-off
3. All emergency equipment must be installed
4. The only available lighting must be that of the emergency lighting system
5. The persons simulating passengers must comprise a representative passenger load, such that
   a) at least 30% must be female
   b) at least 5% must be over 60 years of age, and 30% of these must be female
   c) between 5% and 10% must be evenly distributed under the age of 12
   d) Three dolls must be included to simulate children or less than two years of age.
6. Seat belts must be fastened
7. The cabin crew of the aircraft may be represented by persons having knowledge of the aircraft exits and safety procedures. Such persons must be seated in their normal allocated seats with belts, and shoulder-harness if fitted, securely fastened
8. The simulated passengers will be informed in the usual way of the position of each exit, but must not be briefed about which particular exit to use in a demonstration
9. There must be no rehearsal of the demonstration for the passengers
10. All evacuees must leave the aircraft by the means provided by the aircraft equipment.

Whilst it is clear that a demonstration of an evacuation under standardized conditions forms a necessary part of the certification process, it should not be assumed that such an exercise demonstrates that a genuine evacuation could be performed within the same period of time. The volunteers taking part in the demonstration will be fit and alert; they will have no problems with hearing, or with understanding the language of the instructions given; they will have heeded those instructions; the aircraft will be intact and level; the evacuation will not be conducted in complete darkness or in smoke; no injured passengers will require attention or cause blockage of egress routes; no items of luggage or other articles will obstruct exit routes. Most importantly, the volunteers will not be in fear of their lives and those of their family and friends, and therefore no freezing or panic reactions can be expected.

It is important, of course, that conditions at least as stringent as those of the demonstration should be imposed upon any experimental studies from which data are to be derived concerning passenger behaviour during emergency evacuation. Studies upon samples of people lacking proper proportions of the extremes of the age range, for example, will yield unreliable results. Similarly, one or two rehearsals are likely to lead not only to a speeding-up of the evacuation, but to qualitative changes in individuals' tactics, thus rendering any data obtained quite unusable. Again, major changes to any part of the normal escape route will render simulation invalid. Delays due to the reluctance of passengers to jump onto a slide will, for example, cause blockage in the doorways and the aisles thus rendering it impossible for some passengers to leave their seats and thereby affecting the whole pattern of evacuation behaviour. Hence the absence of a slide from a mock-up can give rise to quite spurious data.

In view of the effects of stress induced by a genuine emergency, the value of data gathered during certification demonstrations and experimental studies must be treated with caution. Some credence might be placed upon rank ordering of performance values resulting from different experimental conditions, but the absolute levels of performance should be regarded as highly optimistic estimates of the levels to be expected under the most favourable conditions during a genuine emergency. The effect of motivational factors upon human performance should not be underrated. The motives of experimental subjects are those of cooperating in a piece of research, incremented perhaps by the possibility of obtaining a small cash reward. In an emergency the motive is the preservation of life.

American investigators carried out an elaborate programme to evaluate proposed configurations of emergency exits for the Douglas DC-8 in order to provide a sound basis for any modifications in the regulatory requirements for exits (R12). Using filmed records and tracking devices whilst their subjects passed through exits of a variety of dimensions, these authors were able to show that transit times were a function of numerous variables relating to the exit geometry and the size, sex, and age of the subjects. Negligible differences were found to result from exit widths within the range 19-24in (483-610mm); this dimension seems unlikely to give rise to problems in respect of typical aircraft skin structures. Escape times increased markedly as the height of the opening was reduced below 50in (1270mm), such that the time was doubled at the minimum height of 26in (660mm) for a Type IV exit. The step up and step down associated with these exits also produced considerable slowing of egress. A step down of 36in (914mm), the maximum permissible in a Type IV exit, increased the egress time by a factor of about 3.

Tests which allowed a comparison of door and window egress times in a DC-8 equipped with ceiling-mounted slides showed that the average time to open a Type I door was 10s and the average time to deploy an escape slide was 34s. Once the slide was deployed, the average evacuation rate was 41 passengers per minute. The average time to open a Type III exit was 12.6s followed by an average of 4.7s for the first person to pass through the exit and thereafter an evacuation rate of 32.8 passengers per minute (T04).

Average exit times from three different exit types based on records of genuine evacuations showed that, after the exit was opened and the slide deployed, twice as many passengers could leave the aircraft in one minute by means of a Type I door (60 passengers) compared with a Type III exit (30 passengers). Comparable figures for a Type A exit were 104 passengers per minute (J13).

The effect of the different exit types on the evacuation of disabled individuals has been investigated using dummies to represent totally incapacitated individuals (B13). It was shown that these could be helped through a Type III exit more quickly than through a door as more time was taken at the door to orient them for the correct position on the slide. However, the investigation did not include movement from the wing to the ground where the likelihood of injury is greater, and, as the investigators pointed out, handling a dummy is not equivalent to handling an incapacitated person. This study also indicated that narrow aisles created a problem for those assisting handicapped individuals to the exit.

Consequent upon an accident to a Boeing 737 in 1985 (A02), the FAA initiated a study of the effect of seat configuration upon the use of a Type III emergency exit (R08). The basic seating configuration comprised three-abreast seats on either side of a 17in (432mm) aisle. Modifications were carried out to achieve the four different arrangements illustrated in Figure 8.4. The removable exit was 508mm (20in) wide and 965mm (38in) high. The lower edge was 457mm (18in) above the floor. The unit weighed 16.3kg (36lb).

The first phase of the study comprised the removal of the exit plug by the person seated adjacent to it. Each of forty experimental subjects performed a single trial having first read an abbreviated passenger briefing card illustrating the method of exit operation. The timed trial began with the sounding of a buzzer, and ended when the experimenter judged that the exit was available for use.

No significant differences were found between the average times taken to remove the exit plugs consequent upon the four different seating configurations. One feature which emerged during the investigation was the importance of disposing the plug following its removal. The investigators commented that "an indiscriminately discarded door may well be more detrimental to safe and rapid egress than many other controllable factors" (p.14).

In the second phase of the study, four groups of thirty-three paid volunteers carried out evacuations from the mock-up by way of the Type III exit. The four exit configurations were used within a balanced-sequence experimental design. The overall average flow

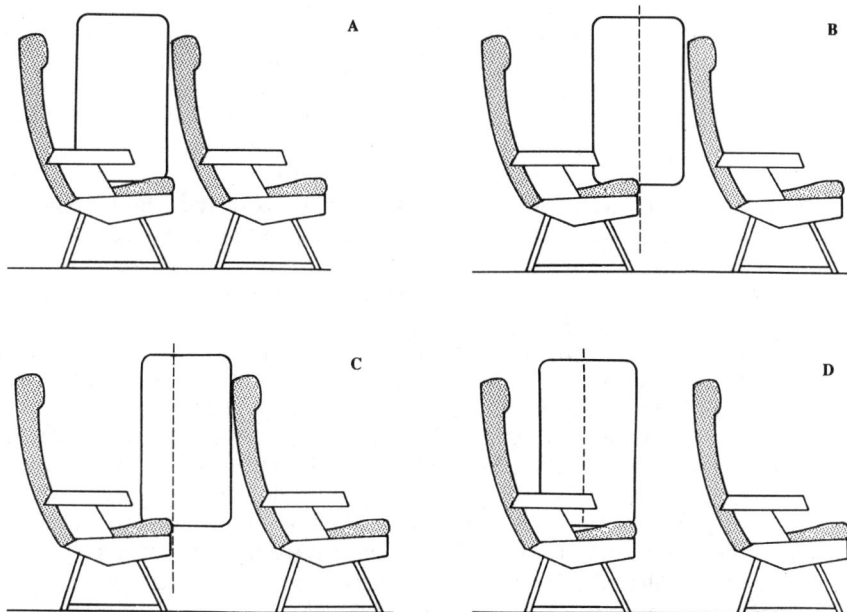

Figure 8.4 Four configurations of seating. A = Pre-1986 FAA and CAA minima with seat backs clear of exit. B = CAA AN79 requirement with 50% of exit width free from obstruction. C = 75% of exit width free from obstruction. D = 32" pitch seat row centralized upon exit but with outboard seat removed.

rates of personnel through the exits are shown in Table 8.4.

Somewhat similar studies carried out in the UK under the auspices of the CAA have been reported (M30A). Seven different seating configurations were employed with widths of clearance between seats varying between 3in (76mm) and 34in (864mm). The configuration requiring the removal of the outboard seat adjacent to the exit was also included. The width of the Type III exit used appears to have been about 30% greater than the statutory minimum. Some of the trials provided the volunteers with the

Table 8.4   Mean flow rates (passengers per minute) for each of the four seat configurations.   The decrement indicates the proportional penalty of sub-optimal arrangements.

Configuration	Mean Flow Rate	Decrement (%)
A	37.0	15.5
B	39.5	9.8
C	41.7	4.8
D	43.8	–

opportunity to obtain additional cash payments were they to be in the first 50% to clear the exit.   Whilst egress times were longer in the case of the minimum seat separation (as in configuration A, Figure 8.4) no clear pattern emerged between the remaining configurations.

Since both the US and the UK studies were carried out using an exit on the left side of the aircraft, no information was available concerning possible differences due to laterality.   More importantly, neither study provided information concerning the effect upon egress time of different methods of disposing of the exit plug following its removal.

Studies of exit times have also been carried out in relation to vehicles other than aircraft.   One such investigation measured the effects of different geometries of emergency doors and windows in order to examine the requirements for the evacuation of public service road vehicles. The size of the openings, the step down, together with the age and sex of the subjects were again shown to have substantial effects upon evacuation times (E05).

A study has also been made of the effects of certain types of physical disability upon mobility within buses (B21).

A simulated emergency evacuation of an aircraft cabin was carried out in the United States some years ago in which attempts were made to reproduce as realistically as possible the conditions pertaining to a genuine emergency involving fire.   The cabin crew used all their skills to encourage the simulated passengers to leave the smoke-filled aircraft with all speed.   Many of these passengers considered the experience to be harrowing, and numerous injuries were sustained as a result of egress by way of the inflatable slides. It seems unlikely that such an expensively realistic experiment will be repeated.

In view of the practical and ethical barriers which prohibit experimental studies of human behaviour in contrived but genuinely dangerous situations, data are obtainable only from the analysis of genuine emergencies or from the limited simulations described above. It must be emphasized that the information obtained from laboratory studies in which certain variables, such as the dimensions of doors and windows, are varied cannot be used directly to predict performance during a disaster. At best, performance might be expected to be degraded by the additional problems and stresses of the genuine event. At worst, disasters might elicit behaviour patterns wholly different from those exhibited in the laboratory, thus giving rise to unpredicted evacuation performance.

Exits and external fire
If there is fire outside an exit, that exit should not be used. Checking for evidence of an external fire is part of the procedure carried out by flight attendants before emergency evacuation. It may be difficult to make an assessment of ground conditions unless the window gives sufficient visibility both downward and outward. To counter this difficulty, the Special Aviation Fire and Explosion Reduction Advisory Committee (SAFER) recommended that heat sensors or fibre optic viewing devices be placed at the doors (S01). Periscopic devices are provided in some aircraft doors.

In the event of a post-crash external fire, breaches in the fuselage allow the fire to enter the cabin. The same effect can result from opening a door. The doors in narrow-bodied aircraft are normally operated manually and may therefore be re-closed. The case is different in wide-bodied aircraft, where the very large doors cannot easily be re-closed. For example, the emergency door-opening mode in a McDonnell Douglas DC-10 is pneumatic and in a Lockheed L-1011 is spring-powered; once the handle is actuated in emergency mode, the door is opened and may not always be closable. The introduction of fire curtains at these doors would reduce the hazard.

ESCAPE SLIDES

Escape slides have developed from relatively primitive fabric chutes to the inflatable slides treated with fire-resistant coating which are currently in use.

Slide evolution
Early escape slides depended for successful implementation on the cooperation of able-bodied male passengers, two of whom were detailed to climb down the fabric chute after it had been attached to the door sill and thrown out of the aircraft. These two passengers were expected to hold the handles at the ends of the

chute, making it taut enough for two more able-bodied male passengers to leave the aircraft and assist in supporting the chute for the remainder of the aircraft's occupants. This method of evacuation, while superior to climbing down a knotted rope, was nevertheless inadequate for a number of reasons. It took time to make the chute operational; evacuees tended to pile up on the ground (B03). There is anecdotal evidence of able-bodied men being unwilling to cooperate, departing from the scene and leaving women to support the ends of the chute.

During the 1950s, the inflatable slide of synthetic material was developed. In the course of the following years, this underwent a series of modifications, including a breakpoint at the bottom to prevent evacuees coming into sudden and damaging contact with the ground and a fabric surface to reduce the possibility of losing inflation from punctures caused by high-heeled shoes, or other sharp objects.

Further developments have resulted in improvements in the fire resistance of slides. By treating the slide with an aluminized coating, the resistance of a slide to the effects of radiant heat from a fire can be significantly increased.

Current design

The exits from current transport aircraft are typically high above the ground. In a wide-bodied jet aircraft, the height of the exit may be 5m (15ft) from the ground, and the top deck of a Boeing 747 about 8m (25ft). When an exit is more than 6ft (1830mm) above the ground, regulations require that it be equipped with "an approved means to assist occupants to reach the ground safely in an emergency" (BCAR D4-3 4.3.1). This takes the form of a self-supporting slide, deployed automatically in current aircraft when the exit opening mechanism is actuated, which must be fully inflated within 10s. In older aircraft, slides are deployed manually. In order to avoid automatic slide deployment during normal use of the door, armed and unarmed modes are available, the former being selected by the crew as soon as the doors are closed and the aircraft is ready to taxi. The doors will be unarmed by the crew when the aircraft again comes to rest. Occasional accidental deployments on the ground have led to expensive delays.

While narrow-bodied aircraft have single slides, wide-bodied aircraft have double slides that can accommodate two people side-by-side. Slides are available at the overwing exits of some aircraft which reduce the likelihood of injury sustained by jumping from the wing to the ground. At the bottom of some slides is a deceleration pad the purpose of which is to decelerate evacuating passengers into a standing position to facilitate rapid movement away from the aircraft.

Adverse conditions for slide use

In the event of an accident, there is no assurance that the aircraft will come to rest in a normal attitude. The slide, consequently, may not assume the intended angle of about 37°-40° to the horizontal. A substantial discrepancy in either direction can lead to difficulties, including the case where the slide fails to reach the ground. This might well occur should the aircraft come to rest with its tail on the ground, in which case the doors forward of the main gear will be unusable.

At angles greater than 45° the speed of sliding increases fairly rapidly. In addition, the passengers are likely to balk at the steep appearance, adding to evacuation time. At around 28°, the speed of sliding is vanishingly small and the passengers must push themselves down, again adding to evacuation time. However, further decreases to below 22° result in the ability to run along the slide (B03).

High winds may cause problems in deploying and using escape slides. The requirement for slides to be capable of operating in winds up to 25mph (11.2m/second) dates from those installed since 1983 but this does not apply retrospectively.

Slide failure

The design of slides must take into account the method of deployment and ensure that this is fail-safe as far as possible. There have been reports of difficulties in deploying slides. They have been known to fail to drop out of the aircraft and have to be pushed out, wasting valuable time. On some occasions, the slides have failed to inflate fully and this reduces their effectiveness in emergency situations.

It is not practical to check the slide and its associated mechanisms on each flight for reasons of time, and therefore cost. Replacing a deployed slide is not something that can be done quickly. It is therefore of major importance that the routine maintenance and repair of these items is carried out scrupulously. An NTSB special report recorded that slide failures occurred frequently, implicating improper installation and maintenance (N07).

Slide damage

Some slides have been destroyed by fire before evacuation could be completed. In 1978, a slide failure in a McDonnell Douglas DC-10 was attributed to radiant heat from a fire located 5m away (N15). In subsequent tests carried out by the FAA, it was found that it took 64s for an aluminized-coated slide at 15ft from an extensive fuel fire to lose initial pressure. This compared with 28s for the uncoated material in current use.

The integrity of the inflatable slides is clearly of considerable importance. At the initial loss of air pressure, the slide has lost its value as an escape mechanism. Sharp items are hazardous to slides and it is for this reason that high-heeled shoes particularly are discouraged.   Broken glass from tax-free bottles can also be a hazard, not only to the slide but also to anyone unfortunate enough to land on it.

Slide–associated injuries

Some injuries associated with slides are those from falls over the sides which have led to the suggestion that slide design should be modified to incorporate high sides.   In one accident, when the attitude of the aircraft caused the forward slides to assume an angle of 68°, some of the passengers using these slides were seriously injured (P03). Injured passengers in their turn are likely to disrupt the flow of evacuees from the aircraft or to cause further injury to others who collide with the injured persons.

EMERGENCY LIGHTING

An emergency lighting system is required in the cabin, the power supply of which is independent of the main lighting system. This must provide illumination for the cabin, for emergency exit areas, for emergency exit marking and for the location of signs.   Exterior illumination must also be provided (BCAR D4-3 4.4).

Smoke

One of the problems associated with a post-crash fire is the effect of smoke on visibility within the cabin.   By obscuring exits, exit signs, aisles, and obstructions such as carry-on baggage, items from the galley, or even bodies, the smoke impedes evacuation and threatens survivability.   It is known that some individuals have passed by exits and perished (S18).

A particular problem arises from the characteristic stratification of smoke, which becomes more opaque as it approaches the ceiling. Illuminated exit signs, typically placed at a high level with the intention that they may be seen by all the occupants in the cabin, are likely to be obscured by smoke when conditions in the lower part of the cabin still permit some vision.

Not only does smoke obscure visual clues, it also has deleterious effects on the eyes which reduce visual effectiveness.   Even quite low concentrations of irritant gases will produce loss of visual acuity and impairment of vision (G14).   An early study on the readability of self-illuminated signs in a smoke-obscured environment showed that, in white smoke, substantial increases in character sizes of the signs resulted in only moderate improvement in readability. The investigators concluded that in extreme conditions of smoke, the question of readability of signs "may well be a moot question unless protective devices such as goggles, face masks or smoke hoods are available" (R06).

Location of emergency lighting
As the introduction of larger or brighter signs was shown to be relatively ineffective in compensating for high smoke densities, a different solution was required. To counter the effect of smoke on the visibility of ceiling-mounted lights and illuminated exit signs, the SAFER report recommended "additional lighting at or below armrest level to provide emergency evacuation guidance and illumination in the relatively clear air found at lower cabin levels" (S01).

A study comparing ceiling-mounted and lower cabin-mounted lighting during evacuation trials showed that evacuation times in laboratory conditions could be reduced by almost one-fifth in a cabin filled with white smoke when emergency lighting and exit locator signs were mounted at or below the midpoint of the cabin, directly illuminating main and cross aisles. It was noted that the individuals taking part in the evacuations in which smoke was layered most heavily in the upper third of the cabin tended to crouch down or stoop to avoid the smoke, and were looking for the exit from just above seatback height. Those individuals who were first down the aisle had the greatest need to see exit locator signs below the layered smoke, as the others followed in their wake. The low-level lights also appeared to reduce disorientation as compared with overhead lights and signs (C07).

Standards for emergency lighting
In 1984, the FAA published new standards for visual guidance in emergency evacuation of the cabin when all sources of cabin lighting more than 4ft (1.2m) above the aisle floor are totally obscured by smoke. This required that within two years, floor-proximity emergency-escape path marking should be introduced in transport category aircraft.

The floor-proximity emergency-escape path marking must enable each passenger to identify visually the emergency escape path along the aisle of the cabin floor after leaving a cabin seat, and readily identify each exit from the emergency escape path by reference only to markings and visual features not more than 4ft above the cabin floor.

In its most simple form, this lighting system is limited to providing route guidance to the nearest door. This is not necessarily the best, or even a viable, escape route. An external fire may, for example, render unusable some or all the doors on one side of the aircraft. The slides on all the forward doors may be unusable should the aircraft adopt a nose-up attitude, in which case the indicated escape path in the forward section will lead to quite the wrong place. One door may have been unusable throughout the flight due, for example, to an unserviceable slide and may have been so placarded. Alternative guidance systems, with more sophisticated levels of adaptability to particular circumstances, create problems of reliability, especially during conditions of emergency involving damage to the aircraft.

The legislated standards do not specify the method by which the visual guidance should be achieved. A study was carried out under the auspices of the FAA of eleven systems including incandescent, fluorescent, electroluminescent and self-illuminated lighting elements at different locations and distributions within the cabin including aisle seat frames, armrests, side-wall panels, aisle floor and overhead baggage racks. As no one system was clearly superior to any other, a performance standard - rather than a particular method of implementation - was adopted in the regulation. The system is required to mark the emergency path to the exits when all sources of illumination above 4ft from the aisle floor are totally obscured by smoke (F03).

External emergency illumination

In addition to internal illumination, the illumination outside the cabin is important if evacuation has to take place in the dark or in other conditions of low visibility. Without adequate visibility, it may not be possible to determine that the evacuation slide has been deployed and inflated adequately, or what the conditions are outside the aircraft. Lack of illumination has caused passengers evacuating over the wing to lose their sense of orientation and fall off the wing on to a hard surface, incurring serious injuries (N07).

Regulations require that exterior emergency lighting should illuminate the path outside overwing exits along the wing to the ground and that each slide should be visible from the emergency exit to which it is attached. The emergency lighting system must be supplied with sufficient energy to provide illumination for at least ten minutes after the emergency landing.

OXYGEN MASKS

The need for oxygen

Pressure in the cabin of contemporary pressurised aircraft is maintained at around 6,500ft and in older jet aircraft at around 8,000ft. If cabin pressure is lost, the several minutes required for an aircraft to lose height may be greater than the time of useful consciousness (see Chapter Seven). The effects on the body of sudden decompression include loss of visual acuity and hearing, and paralysis of the limbs. The reduction of oxygen available to the brain may lead to loss of consciousness, or, in extreme cases, to death. The use of supplemental oxygen is therefore imperative and passengers must don oxygen masks immediately following their presentation. Oxygen may also be required for medical purposes (see p 157).

The legislated requirements both in the UK and the US are quite complex, being dependent upon the type of aircraft, the altitude and duration of the flight. Differing rules apply respectively to flight crew, cabin crew, and passengers. The objective is to ensure that, in the event of a sudden decompression, sufficient oxygen is available to protect the occupants of the aircraft until it becomes possible to maintain an altitude at which no supplementary oxygen is necessary. Even at the maximum safe rate of descent, several minutes might elapse before such an altitude can be achieved. In some parts of the world, the presence of high ground will prohibit an immediate descent to a safe altitude. Certain states allow aircraft to operate with completely inadequate facilities for the provision of oxygen.

Storage of oxygen
Gaseous oxygen may be stored at high pressure in cylinders and distributed throughout the aircraft by means of oxygen lines to each oxygen mask. Alternatively, chemically generated oxygen may be supplied at each passenger position. This has the advantage of eliminating the need to carry oxygen at high pressure with associated weight penalties and, because the ingredients are inert until activated, no additional fire hazard is created. However, once started, the chemical reaction cannot be stopped and the container becomes very hot. Typically, about 2 litres/min of oxygen are available to each passenger.

Oxygen masks may be stored in overhead compartments, or in compartments in the back of the seat in front. There is an automatic release of oxygen masks when air pressure in the cabin reaches 12,500-14,000ft.

Construction of oxygen masks
Numerous designs of oxygen mask are currently in use. A typical example has attached to it a strap and a plastic reservoir bag, the latter attached in its turn to a tube connecting the system to the oxygen supply. The strap and plastic bag are packed inside the mask. For maximum effectiveness, the mask must fit tightly on the face and the purpose of the strap is to provide adjustment for a tight fit and to ensure that the mask stays in place with the passengers' hands free, particularly in the event of loss of consciousness.

The action of pulling the mask to the face is intended to have the effect of dislodging a pin or pulling a cord which causes the oxygen to flow. Some free-flow systems require no such activation.

Problems associated with oxygen masks

There may be failure of the automatic system to release the oxygen mask in conditions of decompression and in these circumstances the compartments may be opened manually by cabin staff.   However, this clearly adds to the total length of time before oxygen becomes available.

The design of the seat-back oxygen compartment has been criticised as requiring excessive passenger involvement and response. "Passengers are reluctant to disturb the neatly packaged system....The presentation...including the generator, the linkages, the piping and the connections, tends to confuse and frighten passengers and, rather than experiment with it, they ignore it" (N09).

Even highly trained individuals have experienced problems in donning masks. Difficulties were observed among these individuals in extricating the reservoir bag and head strap from their folded positions within the mask; the lanyard was found to interfere with efficient mask-donning; limited filling of the bag resulted from the bag being twisted or caught under the strap. A further problem was that of strap slippage down the back of the head which allowed leakage before the strap was re-tightened (B28). The strap is adjusted by pulling the ends, which are short and close to the face, and thus difficult to find and to grasp.

Once the mask is donned, problems may arise if individuals are unsure of what to do, or how hard to pull on the mask in order to start the flow of oxygen.

There may be further difficulties in recognising that the oxygen is flowing to the mask.   The reservoir bag does not expand and contract at each breath and at a cabin altitude of around 14,000ft, the flow rate is less than 1litre/min.   As oxygen is tasteless, colourless and odourless, there is no means of determining whether the steps taken to ensure the flow of oxygen have been effective. This lack of feedback may be very disconcerting. To counter this, an oxygen flow indicator has been incorporated into some masks. This is a device inserted into the tube leading to the reservoir bag which reveals a green cylinder when oxygen is flowing through. Another method of indicating oxygen flow is to separate a small part of the reservoir bag where the tube enters it. This small area is coloured green and fills with oxygen before the oxygen flows through to the larger section. The green section bears the legend "Green inflated - Oxygen OK".

Considerable heat is produced during the chemical generation of oxygen and the metal parts of the compartments may become very hot and thus a hazard to the unwary. Burns on fingers and hands have resulted from touching the metal.

As the aircraft descends to land and supplemental oxygen is no longer necessary because of the lower altitude, the seat-back oxygen compartments must be closed otherwise they would constitute a serious hazard during landing in the same way as would tray-tables if not stowed. Closing these compartments and re-setting the latching mechanism is a task for the flight attendants. Oxygen masks and bags have been damaged and melted from the heat of the generators when being re-packed by passengers.

Effects on communication
The experience of breathing through an oxygen mask is very different from normal breathing. For the latter, muscular effort is used for inhalation whereas with the mask, it is exhalation that demands the effort.

There are effects on speech of both high altitude and hypoxia. Voices sound different because sounds are attenuated and because vowels are more subject to change than consonants. This makes communication difficult. These difficulties are exacerbated by the effects of reduced pressure on equipment such as microphones.

Additional oxygen masks
FAA regulations require that for aircraft certified for altitudes greater than 30,000ft, the total number of oxygen dispensing units and outlets must exceed the number of seats by 10%, and the extra units must be distributed throughout the cabin as uniformly as practicable. Extra oxygen masks may be needed by flight attendants or for people travelling with infants not occupying a separate seat. The allocation of seats at check-in to adults with infants should also take into account the location of the extra oxygen masks.

Portable oxygen bottles
Regulations require that portable oxygen bottles be provided throughout the cabin at flight attendants' stations for use by cabin staff in the event of a decompression or for first aid application. Since some of these bottles may be heavy, they are normally provided with a strap to assist carrying. Colour coding is employed to avoid confusion with other items, particularly fire extinguishers.

Some valves deliver oxygen at a fixed rate, others have two settings such that the supply can be varied, typically in the range of 2-4 litres/min. Flow indications are fitted and must, of course, be checked when the equipment is put to use. The duration of the supply is normally between 30 and 60 minutes. Smoking is prohibited in the vicinity of a therapeutic oxygen supply. On some aircraft, therapeutic oxygen is also available from outlets adjacent to passenger seats.

Maintenance
It is important to recognise that oxygen masks, together with all safety equipment, depend for their effectiveness on adequate

maintenance. That this is not always carried out to the highest standards is evident. In a decompression in a Lockheed L-1011, 20 oxygen compartment doors failed to open and "failure of oxygen compartment doors to open has been noted in mechanical reliability reports" (N09). Reservoir bags have failed to inflate when the sides have stuck together and when the vinyl plastic has become inflexible from exposure to low temperatures. A report concerning the oxygen masks in one aircraft stated that "during role change conversion, a number of passenger service modules were found to contain incorrectly stored oxygen masks. The lanyards were not attached to the hoses, hoses were counterwound with valves at the wrong end of the mask racks and Teflon flaps were not covering the latches. The masks would not have deployed in an emergency" (C17).

Maintenance of oxygen bottles is an important consideration. Oxygen will not be obtainable from bottles where the valve controlling the flow is not fitted correctly.

CABIN FURNISHINGS

Luggage bins

Storage compartments in the cabin must, like seats, be designed to withstand loads specified by regulation. However, "experience has shown that stowage compartments do not meet the intent of the regulation to keep their contents from being hazards by shifting when subjected to the specified forces" (N17).

If an aircraft encounters severe turbulence, or has to make an evasive manoeuvre to avoid collision, or has been forced to land, there is a high probability that the catches on overhead lockers will fail and the contents released on to the heads of the passengers seated below. This probability is reflected in the estimate that, in 78% of survivable accidents, overhead lockers burst open. When catches fail, heavy hand luggage and bottles of tax-free alcohol become "missiles" inflicting injury which may cause serious impediment to effective escape (N17).

The failure of lockers to maintain their integrity has a further consequence and that is the obstruction of aisles and exits by the stored items. Pillows and blankets which are unlikely to cause injury may serve to impede passengers in their attempts to evacuate the aircraft.

A petition submitted to the FAA in 1984 by the Association of Flight Attendants requested rules for measuring and limiting carry-on baggage, and proposed a means of screening carry-on baggage prior to boarding, as required in existing regulations. It proposed that carry-on baggage be permitted if it fitted within a space measuring 9 x 16 x 20in (230 x 406 x 508mm).

Steps have been taken in Australia to allow carry-on baggage provided that it weighs no more than 4kg and can pass a "template" test in respect of linear dimensions. Any other baggage must be carried in the hold.

There is an FAA draft regulation restricting the size of carry-on bags such that they fit under the seat. Reserving the lockers for soft items (coats, blankets) would reduce the probability of injury from items striking passengers and flight attendants when the catches of the lockers fail.

Polymeric materials and combustion

Both "hard furnishings" (interior walls, bulkheads, ceilings) and "soft furnishings" (carpets, curtains, seat cushions) in aircraft utilize the weight advantages, noise-absorbing and hard-wearing qualities of polymeric materials. These materials are also used for counter tops, serving trays and in wall-lining insulation panels. However, the penalties associated with these materials become evident in the event of a fire.

In addition to the usual effects of combustion on oxygen depletion, the ignition of polymeric materials produces smoke and toxic fumes which can at best hamper efforts at evacuation and at worst cause death.

When polymeric materials burn, they produce a thick, hot, dense smoke. This smoke, which tends to accumulate near the ceiling, may be hot enough to burn the skin and to set clothing alight. It reduces visibility by obscuring the light and also by irritating the eyes, which produce tears which in turn further reduce the ability to see.

Toxic gases

There are toxic gases in the smoke, including carbon monoxide which is lethal in quite small quantities; 1% may be effective within one minute in causing death. At sub-fatal levels, carbon monoxide can cause vertigo and impaired judgment, leading to confusion and eventually loss of consciousness as concentrations increase. In addition to carbon monoxide, various forms of plastic produce other toxic gases such as hydrogen cyanide, hydrogen chloride, ammonia, and phosgene. Phosgene may be incapacitating at very low concentrations and with very few breaths (K03). Hydrogen cyanide causes people to breathe more rapidly, thus leading to the ingestion of greater amounts of other toxic gases. As the fire burns, the oxygen in the air is depleted.

In the event of an in-flight fire, survival is likely to be more endangered by the effects of toxic gases than by burning. In 1973, a Boeing 707 made an emergency landing near Paris only ten minutes after the presence of a cabin fire was detected. The fuselage was intact and all the exits were operable. However, only the occupants of the cockpit and two stewards forward in the cabin were able to evacuate the aircraft unaided and only 11 survived the accident. Carbon monoxide poisoning was considered to have rendered the other occupants unconscious by the time the aircraft landed and to be the cause of death in more than three-quarters of the cases, the remainder dying from the effects of other toxic gases (I06).

If an impact-survivable accident is followed by fire, this fire may itself prevent or hinder successful evacuation of the aircraft. Where the fire is not immediately life-threatening, however, then the influence of smoke and toxic gases from the combustion of interior materials on evacuation may be significant. Smoke can cause serious difficulties for occupants attempting to carry out visual tasks such as reading instructions or finding the way to an exit. It can also produce disorientation such that individuals may not know in which direction to go to find the exit. Evacuation may be further hindered by the effects of smoke on the respiratory system, making breathing painful and difficult, and possibly causing uncontrolled coughing reflexes.

A post-crash fire typically starts as a result of the ignition of fuel mist arising from spilled fuel outside the aircraft. This fire will attack the fuselage and burn through into the cabin typically within 40-120s, the faster time if the fuselage is damaged or doors have been opened (K03). Once the fire has penetrated into the cabin, smoke and toxic gases are generated and conditions deteriorate rapidly.

The polyurethane foam with which seat cushions are upholstered has been identified as an important source of toxic gases when ignited. This foam is difficult to ignite at low temperatures but when ignited by high temperatures (400°C) and in bulk quantities, polyurethane foam produces thick smoke, carbon monoxide, hydrogen cyanide, phosgene and ammonia.

Reduction of fire and its effects

A number of regulatory steps have been taken with the aim of reducing the possibility of fire in the cabin and, should a fire break out, of minimising its effects.

Recent regulations require that the toilet compartment be fitted with a smoke detector and that the receptacles for waste items be equipped with automatic fire extinguishers. One source of these fires derives from passengers smoking illicitly in the toilet compartment, or disposing of a glowing cigarette into the waste receptacle upon entering. Passengers would be well-advised to co-operate with the crew in attempting to enforce the prohibition of smoking in this compartment.

In the event that an in-flight fire has started, the mandatory provision of protective breathing equipment (smoke hoods) for cabin crew will facilitate their efforts in fighting the fire. This measure is supported by the requirement for hands-on training in fighting fires.

The requirement for a fire-blocking layer to reduce the flammability of foam seat cushions was enforced by the FAA and the CAA in 1987. This modification has the effect of delaying the onset of ignition and reducing the spread of flame thus providing 40s more survival time.

Following upon extensive research studies, both the CAA and FAA promulgated mandatory changes during 1987 requiring more severe test standards of flammability of cabin materials (C15; FAR 25.853). These standards are based upon the assumption that flammability and the emission of toxic gases are correlated and that the severe hazard from toxic emissions occurs as a result of flashover. However, doubt has been cast upon these assumptions by the results of recent tests of fire-blocking cushions on seats conducted by the FAA in 1987. These showed that although flashover was delayed to 240s after the initiation of the fire compared with 135s where fire-blocking material was not used, toxic gases in greater than fatal concentrations were detected before the onset of flashover. Consequently, it has been suggested that the level of toxic emission should be employed as the criterion of fire-blocking materials (A02).

Low-level illumination and tactile markers are more effective in conditions of smoke in guiding passengers to exits, and regulations requiring such illumination are now in place.

During in-flight fire, the role of ventilation is important. This has been misunderstood both by flight crews and cabin staff on some occasions when, instead of maximizing ventilation to remove the smoke from the cabin, air conditioning has been closed down in the mistaken belief that it fans the flames (N20).

The use of oxygen masks in in-flight fire has been controversial. One report claims that "many pilots owe their lives to... disciplined use of their oxygen mask in a smoke- and toxic gas-filled cockpit" (B15). However, the use of supplemental oxygen in a fire environment is considered to add to the existing hazards (B16). In the absence of decompression, passenger oxygen masks will not deliver oxygen although they have been manually deployed in cases of in-flight fire, with reported benefit to the passengers (see p. 114).

Another controversial countermeasure is that of smoke hoods (see p.168). It is considered that the introduction of smoke hoods would permit the use of improved fire retardant materials for cabin furnishings which would "otherwise result in an unacceptable toxic level in the cabin air" (B15). If these materials were to be used and if smoke hoods were, for various reasons, not worn by occupants, then the fatalities from toxic fumes would be likely to rise.

FIRE EXTINGUISHERS

Fire extinguishers are used to combat an outbreak of fire during flight. Regulations require that each separate compartment within an aircraft be equipped with a fire extinguisher of an approved type. More extinguishers, up to a maximum of three, will be required in the cabin as the number of passenger seats increases. Large aircraft typically carry a dozen or more extinguishers distributed throughout the aircraft.

The active agents of fire extinguishers used in aircraft are water, carbon dioxide, potassium bicarbonate (dry chemical), or halon. Each has advantages and disadvantages (B17). A major disadvantage with the use of water is in the method of operation in which a stream of water projected on to the fire is too fine to be effective. It should not be used on burning fluids or upon electrical equipment, but otherwise is suitable for most purposes within the cabin. Carbon dioxide extinguishers are heavy and awkward to use and the carbon dioxide may have deleterious effects on electronic equipment. When dry chemical extinguishers are used, a cloud of opaque powder is ejected which severely reduces visibility. Extinguishers using halon are generally regarded as most effective in combating fires of all kinds and halon is not a hazard to electrical equipment. For these reasons, the FAA has ruled that at least two extinguishers containing Halon 1211 as the extinguishing agent must be carried. However, the disadvantage of halon is that it produces toxic gases during decomposition in a fire and should not be used on fires which can be extinguished using water.

It is important that the design of a fire extinguisher should take proper account of the characteristics of the user. It should not be unduly heavy to carry or difficult to operate, particularly in the confined spaces on board where fires might start. The contents of the extinguisher should be clearly denoted by the colour of the container.

Hands-on training in the use of the range of extinguishers likely to be used on board is also important in combating in-flight fires. Small fires in passenger seats, however, are usually dealt with swiftly and effectively by the use of a blanket together with water, often from the service cart.

EQUIPMENT FOR SURVIVAL IN WATER

Since the advent of jet aircraft, the probability of ditching, that is a planned water landing, is very small although a few examples have occurred in recent years. Experience has shown that survival rates can be quite high provided that there is time for proper preparation and that control of the aircraft can be maintained.   In contrast, a crash into water immediately after take-off or during an approach to land can bring catastrophic results.

To safeguard against the consequences of ditching, aircraft must carry either flotation cushions (in the US) or life jackets. For flights more than 50 nautical miles from the coast, the aircraft must, in addition, carry life rafts.   Inflatable slides may serve also as life rafts.   Sufficient raft accommodation must be provided to ensure that there is space for all the occupants of the aircraft even allowing that one raft, of the largest type carried, is lost.   Each raft must be provided with a locator light, a pyrotechnic signal, an emergency locator radio transmitter, and a survival kit.

Seat flotation cushions

Though all aircraft seat cushions are likely to float because they are made of polyurethene foam, they will not necessarily remain buoyant for more than about fifteen minutes. Some cushions in use in the US, designed specifically as flotation cushions, are constructed around a block of closed-cell polyethylene which does not absorb water. These flotation cushions are equipped with straps or loops to facilitate their being grasped in the water.

A flotation cushion is required by regulation to provide, when completely immersed in water, a buoyancy of 14lb (6.4kg) for an 8-hour period. Since the human body has a buoyancy close to zero, the cushion when correctly used will, in the majority of cases, allow the head to be kept clear of the surface.   The buoyancy of the cushion is affected by the density of the water, being lower in warm, fresh water than in cold, salt water.

Flotation cushions are more readily accessible in emergencies than life jackets and do not have to be donned. The FAA has stated that life jackets "provide no added measure of protection from hypothermia or the effects of hypothermia over flotation cushions" (F05). However, flotation cushions have disadvantages in relation to life jackets which make them less effective as survival aids. Within the aircraft, they are bulky items to carry and they must be held securely throughout the evacuation. Once in the water, a firm grip must be retained upon the cushion or it will float away. Maintaining a grip will become increasingly difficult as continued exposure to cold water numbs the arms and hands. As exposure time increases, loss of consciousness is likely to result and the grip on the cushion will be completely lost.

Life jackets

Current regulations require that life jackets provide a minimum buoyancy of 35lb (16kg) in 21°C fresh water, that instructions for donning must be printed so that the individual can read them with the jacket on, and that an adult can, after watching a demonstration of the kind provided by flight attendants, don the jacket within 15s.   However, these regulations refer only to life jackets manufactured after the beginning of January 1985 and do not prohibit the use of older life jackets.   These older life jackets must have a buoyancy of 20lbs and the instructions may be printed on the back of the jacket or on the front, upside-down.   Life jackets must be capable of turning an unconscious person from a face-down to a face-up position in a short space of time.

There are life jackets designed for young children less than 16kgs (35lb) in weight, for older children between 16 and 41kgs (90lb) and for adults over 41kg.   Life cots for very small babies are also available.   Life jackets may be fastened in a number of different ways.   Some use snaps and buckles, others use tapes which are tied around the body.   All the fastenings of whatever kind must be adjusted by the wearer to give a tight fit to the life jacket otherwise it is liable to be dragged off when the individual jumps into the water.

Each life jacket is equipped with a light so that survivors may be located and with a means of inflating the jacket by blowing if this does not take place automatically when either of two tabs is pulled.

Experience has revealed numerous problems associated with life jackets.   These include the way in which they are accessed from storage, the instructions for donning (see p.187) and the tendency for passengers to steal them from the aircraft for use on boats.

Life jackets may be stored under the passenger seat, in overhead compartments, or in the backs of seats.   Many problems have been shown to be associated with storage of life jackets under the passenger seat.   The first one is the difficulty of gaining access to the life jacket, which increases as seat pitch decreases.   Where seat pitch is as little as 760mm (30in), it will be impossible for some passengers to reach down far enough to gain access to the area beneath the seat.   For the same reason, it may be difficult to ascertain in advance of an emergency that a life jacket is in place. Secondly, after reaching into the storage area, it may be difficult to remove the life jacket from its niche. Some of the passengers preparing to ditch in a Douglas DC-9 had to go on to their hands and knees to release the strap keeping the life jacket storage pocket closed.   Thirdly, opening the plastic cover containing the life jacket may present problems.   One passenger in the DC-9 used his pocket knife for the purpose (N04).   However, many passengers are unlikely to be equipped with this means of solving the problem and so it is important for the cover to be designed for easy opening to avoid the loss of valuable time.   Life jackets stored in the backs of seats are more accessible.   This storage area, in common with overhead compartments, has the advantage of being less affected by the entry of water into the cabin and by being blocked by stowed luggage, either of which is likely to compound the difficulties associated with retrieval from underneath the seat.

It should be noted that there is very little space available for donning life jackets in seats with a typical tourist-class pitch.

Life rafts

Rescue from the ocean is not likely to take place quickly. Adverse weather conditions may delay airborne rescue while at the same time causing additional difficulties to survivors in choppy seas. After the ditching of a Douglas DC-9 in 1970, thirty miles off the coast of Florida, ninety minutes elapsed before a helicopter arrived and was able to start winching survivors out of the water and three hours passed before the last survivor was rescued (N04).

Submersion in cold water will result in a loss of body heat, eventually resulting in hypothermia. Conversely, body heat may be conserved if the individual is able to get out of the water. Survival after ditching is therefore likely to be considerably enhanced if life rafts can be utilised.

Life rafts in some aircraft are stored in containers near the exits from which they will be deployed; some have to be lowered from overhead compartments. They are sufficiently heavy to require more than one person to carry them. British and US regulations require that rafts should be "easily accessible" and their installation should "not likely to be either an undue hindrance to or unduly hindered by the occupants vacating the aeroplane" (BCAR D6-6; FAR 121.339).

At the exit, rafts are first attached to the aircraft and then inflated when they are in the water. Handles and lines are provided for passengers to use to climb into the rafts from the water. In the ditching referred to above, an inflatable slide used as a raft was found to have insufficient handles for all the survivors to hold. When the life raft has its complement of passengers, the canopy must be erected to provide added protection against excessive sun, wind and precipitation.

One raft in each aircraft must be equipped with an emergency locatory transmitter (ELT) which is powered by a water-activated battery and which automatically transmits signals to enable potential rescuers to locate survivors. All rafts are required to carry a flare.

On wide-bodied aircraft, the inflatable slides are normally slide/rafts which inflate automatically when the door is opened and are disconnected from the aircraft by operating a handle. Canopies on some slide/rafts are partially inflated when the slide is inflated.

There are clearly advantages in slide/rafts as some of the actions necessary to activate life rafts are eliminated. A life raft must be taken from its storage compartment, carried to the exit, secured, launched and inflated. This is not only more time-consuming than deploying a slide/raft, it is also more prone to problems. In the Florida ditching, two members of the crew spent much of the time available for preparation in removing the life raft in its carrying bag from its storage position in the coat closet and moving it into the galley area. They then tried to find the lanyard (used for tie-down and inflation) inside the container. However, the life raft inflated inside the passenger cabin, blocking the exit and trapping the foot of a third member of the crew (N04).

MECHANISMS OF COMMUNICATION

The oral briefings before and during the flight are typically given by means of the aircraft public address (PA) system.

It is important, in the event of an emergency, that the cabin crew can communicate quickly and clearly to all the passengers on board. The public address system, powered by the aircraft's power supply, provides the normal means of achieving this.

Different sound levels may be available on the PA system for use in relatively quiet conditions on the ground and for use in conditions of engine noise. In those emergency situations where behaviour is uncontrolled and disorganized, the PA should be capable of producing sound levels which allow announcements to be heard over the shouts and screams of passengers.

The importance of the PA system is recognised by FAA regulations which require that it is one of the last systems to be lost in the event of an emergency. A greater safeguard would be the provision of an independent source of power for the PA system and this is under consideration by the FAA as a possible requirement.

In the event of failure of the PA system, then an alternative method of communicating is by means of a megaphone. These should be stored near doors and within reach of the shortest cabin crew member. A report by the NTSB in 1974 pointed out that megaphones were rarely used as they were often stowed in overhead racks where they were not within reach of flight attendants while seated. In order to obtain a megaphone, they would have to abandon their assigned duty stations. It is also the case that megaphones may be overlooked if they are stored in inconspicuous and inaccessible places (N07).

It was noted in an accident report that some of the emergency equipment, which included two megaphones, was stored in overhead bins in the cabin and not near the cabin crew stations. Consequently, in an emergency evacuation the cabin crew may find it impossible to reach this equipment as passengers move towards the exit (A02).

Some aircraft are fitted with an evacuation alarm activator on the flight-deck and at the purser's station by means of which the order to evacuate the aircraft can be transmitted to cabin crew.

FIRST AID AND MEDICAL EQUIPMENT

Aircraft will carry at least one First Aid kit and may be required to carry several according to the number of passenger seats fitted (Table 8.5). The contents of each kit, as specified for example in Appendix A of FAR 121, include a variety of bandages, tape, antiseptic dressings, splints, and ointments. The availability of therapeutic oxygen bottles has been noted in a preceding section (p.157).

In addition, it is likely that an Emergency Medical Kit will be carried. Such a kit will probably contain several basic instruments and a variety of drugs required for life support during a medical emergency.

Table 8.5 The number of First Aid kits to be carried by passenger aircraft (FAR 121.309).

Kit	Passengers
1	0 – 50
2	51 – 150
3	151 – 250
4	250 +

A new medical kit introduced by British Airways during 1989 serves as an example. This is arranged in four layers. The first contains the basic items for first aid. The second tray contains such injectable drugs as gluagon, diazepam, adrenaline, aminophylline, and opiates. At the next level are cardiac drugs including isoprenaline, lignocaine, and calcium chloride. Finally, catheters, syringes, a stethoscope and a sphygmomanometer are included (G06).

Different opinions have been expressed by medical personnel concerning the range of equipment which should be available. Defibrillators, for example, have been advocated, but the competence of an itinerant doctor to deal with such equipment has been called into question. A scheme for monitoring cardiac function via satellite telemetry to a ground station has been proposed (P09).

SMOKE HOODS

The purpose of smoke hoods is to "expand the available escape window" by protecting the lungs from noxious fumes and the eyes from irritating smoke (B15). Smoke hoods are not presently part of the safety equipment for passengers on board transport aircraft. They are, however, mandatory for cabin crew.

The need for respiratory support
In-flight fires are typically associated with smouldering and the hazards confronting passengers centre mainly on the inhalation of hot, toxic smoke and fumes. In addition, the process of combustion is likely to cause a reduction in the proportion of oxygen in the air available for breathing. Because passengers must remain seated while in flight, there is little requirement for unimpeded vision, though considerable discomfort will be experienced in the eyes from the effects of irritant gases. What is needed for life support is a "safe, breathable supply of oxygen" (B15). This could in principle be provided by the aircraft's stored oxygen.

A post-crash fire is likely to be a fuel fire and therefore will exhibit characteristics different from an in-flight cabin fire. In contrast to the in-flight cabin fire, passengers must use considerable energy to evacuate the aircraft as quickly as possible. There is the possibility of flashover and of conditions within the cabin quickly becoming very hostile. The time available for emergency evacuation will not be more than five minutes and may be a great deal less. It should be noted that hot, toxic smoke causes problems not only during evacuation but, in the event of survival, may contribute to serious damage to the lungs.

Early attempts at regulation
The value of a simple smoke hood in providing fume protection for evacuation purposes during a post-crash fire was demonstrated in the laboratory in 1967 (M10). The FAA proposed a rule in 1969 that smoke hoods should be provided for emergency use by passengers and crew-members. However, this was withdrawn in 1970 because of the "resulting criticism concerning excessive cost, pilferage, liability, questions concerning passenger acceptability, hazards caused by delay in donning the device or its improper use" (S29). It can be seen from this list that the only criticism relevant to the smoke hood itself was that of the hazards caused by delay in donning it.

A single solution
In 1979, it was proposed by an international committee of experts that a multi-function hood should be designed to fulfil the respiratory needs of passengers during a decompression, during an

Figure 8.5 One design of a smoke hood for cabin attendants.

in-flight fire, and during emergency evacuation in conditions of a post-crash fire (B15). This would have an independent air supply, it would be portable, and on disconnection from the aircraft emergency oxygen system would provide a portable 3-5 minute air supply reservoir.

A study to examine the potential for the development of such a single device, based on the use of oxygen masks and simple smoke hoods, was carried out (S29). Considering the results of this investigation, it would appear that a search for a single solution to three different problems may be mistaken. While the evidence favoured a single device for decompression and the effects of in-flight fire, the problems associated with post-crash evacuation posed problems of a different order.

Supplemental oxygen supplied by means of a mask, as it is at present but without the necessity for decompression to activate the supply, might provide a minimal solution to counteract the effects of both decompression and of in-flight fire. Covering the whole head in order to shield the eyes from smoke is not necessary for sedentary occupants. In sudden decompression, the oxygen may be required for several minutes while the aircraft loses height. In an in-flight fire, the period may be much longer and in these conditions, the life support system attached to a reservoir in the aircraft rather than an independent smoke hood is a more acceptable design solution.

For mobile occupants evacuating the aircraft during a post-crash fire, protection of the eyes from smoke and a supply of breathable clean air for a relatively short period are the essential features.

Design of smoke hoods
A smoke hood at its simplest is a plastic bag pulled over the head, trapping enough clean air for the individual to breath while making a speedy evacuation from the cabin. The smoke hoods used in the experimental programmes in the 1960s were an elaboration of this concept, culminating in the Sheldahl Type S hood which was equipped with a septal neck seal and an aluminized top. These were effective in the laboratory for about 120s before the build-up of carbon dioxide and the depletion of oxygen reached critical levels.

Smoke hood design has evolved considerably since this time because although smoke hoods have not been approved for the use of passengers in aircraft, they have been used on the ground in a number of applications. Sophisticated designs of smoke hoods fall into two groups. There are filter hoods which filter out the noxious gases in the ambient air and may also be provided with catalytic converters to convert carbon monoxide to carbon dioxide. There are hoods equipped with a supply of air or oxygen or a mixture of the two, and provide a closed-loop system by absorbing the exhaled carbon dioxide.

After 1985
In the wake of the 1985 Manchester aircraft fire disaster, much argument ensued about the effectiveness of smoke hoods and their value in reducing the death toll in the event of an aircraft fire. In a discussion document in 1986, the CAA set out the advantages and disadvantages of smoke hoods (C14). The CAA considered that the advantages of a smoke hood were that it provided an extension of the time for which the cabin can be regarded as habitable in the presence of smoke; that it would reduce panic, thereby permitting

a more orderly and rapid evacuation; that it would reduce the risk of an individual being overcome by smoke and collapsing in the aisles and thus causing obstruction; and that it provided eye protection. With the exception of the claim that smoke hoods would reduce panic, there can be little disagreement that smoke hoods offer these advantages.

The disadvantages of smoke hoods were considered to be the time taken to don them resulting in evacuation delay, the false sense of security engendered, and reluctance on the part of passengers to put a plastic bag over their heads. In addition, condensation inside the hood would restrict vision and cause difficulties in communication because of the crackling noise generated by head movements. Problems of storage and accessibility were also factors to be considered, as were problems of pilfering. An additional problem not mentioned in the CAA discussion is the requirement for passenger briefing concerning the use of smoke hoods. Other disadvantages related specifically to filter hoods on the grounds that they do not provide against oxygen depletion, and they are likely to increase the heat of the hot gases drawn into the filter by the action of the catalyst converting carbon monoxide to carbon dioxide.

In January 1988, regulations were announced by the UK CAA that required all civilian aircraft operators to provide breathing apparatus for every member of the cabin crew from January 1990. However, the CAA decided against requiring smoke hoods for passengers on the grounds that none of the types currently available met its required safety standards, including the ability to cope with all the hazardous gases which may be present in a cabin fire. It is questionable whether this is a necessary feature of a smoke hood.

The standards adopted by the CAA were criticised as excessive by those organizations (Parliamentary Advisory Council for Transport Safety, Air Transport Users' Committee, Consumers' Association) pressing for smoke hoods in the passenger cabin. The standards applied to Type I hoods, which specified the requirements for protection during an uncontained in-flight fire followed by an emergency landing and subsequent ground evacuation, and Type II hoods, which specified the requirements for protection in a post-crash or ground emergency evacuation involving fire and smoke.

For both types, there was a requirement that the smoke hood should be capable of being put into effective use by the wearer within 10s of its need being recognized. The Type I hood was required to be effective at pressure altitudes between sea level and 10,000ft for 20 minutes at minimum respiratory rates. Within this period there should be protection provided during a five minute period for the high respiratory demands associated with an emergency evacuation. The Type II hood was intended to provide protection for five minutes at sea-level conditions during which high respiratory rates were assumed. Smoke hood trials carried out by the UK AAIB showed that smoke hoods designed for passenger use could satisfy these requirements (A02).

A simple solution

The cheapest design solution is the Type S hood with its elastic heat-resistant polyurethane seal fitting closely around the neck, requiring no manual adjustment.   Given that the maximum acceptable level of carbon dioxide necessary for movement is 8%, this hood was shown to reach such a concentration when subjected to high temperatures and with an active wearer in just over 120s (M14).

Tests demonstrated that this hood provided users with adequate protection from the respiratory effects of a toxic environment and that even in dense, black smoke, users were able to travel distances which would usually exceed those required to reach the emergency exits in aircraft (M12).   They were also able to execute a large number of switching operations, suggesting that the ability to manipulate exits was not likely to be affected (M12).   Wearing the hood did not interfere with hearing (T06).   However, visual acuity in emergency illumination was significantly affected and an increase in the level of this lighting by 25%-30% was recommended (L01). Since this study was undertaken, however, new regulations concerning emergency lighting have come into operation.

In spite of these positive features, the main disadvantage of this hood is the short period for which it is likely to be effective.

The experimental programme led to the conclusion that "no hood designed to meet important criteria of accessibility and economy of storage can be expected to provide absolute protection and life-support for indefinite periods" (M14).

Estimates about the number of lives saved by the use of smoke hoods are based on the assumption that these are donned correctly and at the most appropriate time.   Such assumptions are very questionable.   While there are many problems associated with the Hardware aspects of smoke hoods, perhaps the most intractable are those arising from the Software interface.   These are discussed below in Chapter Nine.

REVIEW

Design decisions inevitably involve compromise.   Such features as appearance, reliability, safety, convenience, maintainability, and cost will influence the form of an item of Hardware selected for a particular purpose.   The penalties associated with space and weight have greater influence upon decisions concerning aircraft equipment than is typical elsewhere.   The need to satisfy their customers' regulatory authorities is another influence upon the design decisions of aircraft manufacturers.

In addition to the devices fitted to the aircraft, and the life jackets provided for each passenger, a large transport aircraft will carry up to 30 different types of portable safety equipment. Many of these will be replicated making a total of about 250 separate items. Many suppliers are involved in the design and production of this Hardware, with the result that there exists a vast amount of expertise in the engineering aspects of safety equipment. Comparatively few suppliers, however, make any systematic use of HF in the design of their products with the consequence that problems appear at the H-L interfaces.

One facet of this problem, which is familiar both within and outside the aviation industry, is the lack of a clear understanding concerning the location of responsibility for the H-L interface. There is a tendency for each supplier to take the view that it is the responsibility of customers to define their requirements. The end-user, that is the airline, might feel entitled to believe that the aircraft manufacturer should supply a product suited in every way to its intended use. Were such the case, the requirements of people using equipment would receive the equivalent attention paid to the properties of materials or the quality of engineering.

Basically, this problem is one of specification, and can best be resolved by the end-user acquiring the expertise to define his requirements with respect to the H-L interface as precisely as he specifies other parameters of the product. In recent years, there has been some progress in this direction, much of it taking place by way of the leadership provided by the defence industry. Consumer litigation following accidents has provided a further fillip.

In some respects, the pace of progress in the direction of safety can seem exceedingly slow. Seat tie-down serves as one example; many years have passed since discrepancies between human acceleration tolerances and the strength of support and restraint mechanisms were first described. Similarly, the use of furnishing materials which produce lethally poisonous fumes on combustion has persisted long after such characteristics were well established.

Numerous injuries are caused by sharp, hard projecting surfaces and by badly-sited structures within the cabin. Many of these injuries could be eliminated by good design methodologies, often referred to by the somewhat negative term "delethalization".

Not all HF problems owe their origin to short-comings in design. During normal operation, accidents occur in the galley or with service carts as a result of poor maintenance standards. During emergencies, doors may jam, slides may fail to deploy, safety equipment may be missing. HF is concerned with such aspects of the management of H-L interfaces, in addition to their design.

9 Software in Emergencies

> One of the principal benefits of Software as a resource for use
> during emergencies is the preparedness which it bestows. Rules
> will have eliminated certain sources of danger; the crew will
> possess a repertoire of relevant knowledge and skills; passengers
> will have been briefed; procedures developed in anticipation of
> of such an emergency will have been rehearsed. The L-S
> interface, however, demands considerable caution, and is
> deserving of more attention than is customary

THE SCOPE OF SOFTWARE

Software is the collective term which refers to all the laws, rules,
regulations, orders, standard operating procedures, customs,
conventions, and habitual practices which define and describe the
ways in which a system operates. It is convenient to divide this
Software into two parts. The first comprises items which are
formally enacted and prescribes the ways in which the system
should behave. The second part includes the description of the way
the system does behave.

ICAO, the International Civil Aviation Organization, is the
principal international agency, established under the auspices of the
United Nations, concerned with aspects of procedural standardization
within civil aviation. ICAO communicates with the government of
each contracting state which, in turn, is responsible for its own
national organization. Each state is expected either to comply with
ICAO decisions or to give formal notice of its intention to defer or
avoid compliance. IATA, the International Air Transport Association, is a federation of international airlines which also disseminates
information and coordinates discussion on technical and commercial
matters amongst the member airlines. Many of the standards
adopted within civil aviation originate within the various committees
of the SAE, the Society of Automotive Engineers. The Flight Safety

174

Foundation is an organization devoted to the collation and dissemination of safety information by means of international meetings and bulletins. Numerous organizations operating at national or regional level are similarly concerned with aspects of operational software.

The onus of mandatory rule-making falls, of course, upon the legislative bodies of each sovereign state. It is common practice for much of the detail involved in this process to be delegated by the governing body to a specialist agency. Most of the regulations referred to in this volume are those of such agencies in the UK and the USA respectively, namely the Civil Aviation Authority and the Federal Aviation Administration. With respect to the aircraft cabin, these regulations cover the safety aspects of the design of the cabin together with its furnishings and equipment, the licensing and training of cabin staff, and the procedures to be followed in a variety of circumstances and phases of flight.

The formal Software is thus voluminous and complex, and the amount of documentation involved in generating, disseminating, and modifying it inevitably assumes sizeable proportions.

An understanding of cabin procedures demands a good deal more than a knowledge of mandatory rules. Each airline has its policies and styles of operation determined by a variety of historical, geographic, cultural, and economic factors. These are induced during the training period and maintained by supervision and inspection. Further differences can be attributed to the styles adopted by individual senior cabin staff.

Some activities in the cabin will, inevitably, be performed in such a way that a violation of the mandatory rules is involved. Two types of inspection procedure may reveal such violations. The former of these involves a person appointed either by the regulatory authority or by the airline itself whose duty it is to ensure conformity. Corrective, or even punitive, action might follow the reporting of a violation. In the second type of inspection, as exemplified by a HF audit, the objective is to establish the extent to which cabin operation approaches an optimal level of safety and efficiency. In this case, the observance of a violation might well trigger an examination of the cause of the violation which could, in turn, lead to changes in the prescribed methods of system operation. Whilst it is obvious that there is a role for inspection procedures oriented primarily towards the upholding of mandatory standards, it must also be recognised that a system governed wholly by the notion of conformity is one that can neither adapt nor improve.

RULES

Passengers have certain obligations placed upon them when

travelling by air. Compliance with the rules, some of which take
effect even before passengers board the aircraft, will help to ensure
a safe flight.   Some rules originate with the regulatory authorities,
others are rules of particular airlines.

Alcohol
It is an offence for a passenger to be intoxicated on board an
aircraft.   Such a person is a hazard to himself and others in the
event of an emergency.   In addition, such a passenger can be a
nuisance during a normal flight.   Passengers attempting to board
whilst intoxicated should be forbidden to enter the aircraft, but
there is a clear problem here in terms of both diagnosis and public
relations.   In 1988, it was considered worthy of press report when a
group of operators attempted strictly to enforce this rule.   A
further problem arises when alcohol is served to passengers on board
if it is apparent that this will lead to intoxication in some cases.
The dilemma thus presented to cabin staff serves as an example of
the coordination required between the training schedules devoted to
service and to flight safety.

Dangerous Goods
There are many items which must not be carried on board as they
may cause damage to the aircraft or expose the aircraft and its
passengers to danger.   Airline tickets display a list of some of the
more common items which are forbidden such as compressed gases,
explosives, and flammable materials.   Additional information is
available in the form of leaflets published by some airlines (B20).
However, little prominence is given to the danger of certain goods,
and the evidence suggests the public is not well-informed.   It may
not be immediately obvious, for example, why thermometers and
barometers are prohibited from aircraft but, if these are broken and
the mercury spills, considerable damage can be inflicted on the
fuselage.   With increased ownership by North Europeans of second
homes in Mediterranean countries, there is a greater likelihood of
"do-it-yourself" goods being transported which may include paint, gas
cylinders, weed-killer, all of which are banned.
   One difficulty relating to the rules about the carriage of
Dangerous Goods is that, while certain items are dangerous, they
may nevertheless be carried in small quantities for the personal
convenience of the passengers.   Matches and cigarette lighters,
together with aerosols for hairsprays and other toiletries are
examples of exceptions to the general rule.   However, the
consequence of making exceptions is that the rules become less
clear-cut and there may be confusion about the limits of the
exceptions.   Reports of in-flight fires caused by matches in
handbags underline the seriousness of the problem.   In one
incident, the smell of smoke prompted a flight attendant to examine

her handbag in which she discovered a smouldering book of matches.
Before one take-off, the smell of gas led flight attendants to
examine some baggage in the rear of the aircraft in which they
discovered a plastic bag containing three boxes of safety matches
each individually wrapped in a plastic bag.   One of these had
ignited and all the contents burned.   Also in the baggage were
cans of camping gas, books of matches, and a cigarette lighter.

Seat allocation

The allocation of seats at check-in is influenced (where smoking is
not totally prohibited on board the aircraft) by the request for
accommodation in either the smoking or no-smoking section of the
aircraft.   However, there are conventions in many airlines governing
the allocation of seats to children and to disabled passengers which
prohibit them from sitting in rows adjacent to emergency exits.
When allocation of seats at check-in has resulted in children or
disabled passengers sitting in these positions, it is the responsibility
of the flight attendant to re-allocate seats.   In practice, it is likely
that passengers will be requested to move to another seat during
the periods of take-off and landing, and, if time allows, during
preparation for an emergency landing.   However, the conjunction of
an uncooperative passenger and an unassertive flight attendant could
have serious consequences in the event of an unplanned emergency
evacuation.

The solution to the problem of allocating particular seats to
physically handicapped passengers is not, however, simply a matter
of avoiding exit rows. There is also the question of the distance of
the disabled passenger's seat row from the exit.   This has relevance
both for non-emergency and emergency conditions.   If near to the
exit, the disabled passenger has less distance to cover when
boarding and deplaning but there remains the possibility that, during
an emergency evacuation, the speed of able-bodied passengers would
be compromised. If far from the exit, the disabled person has
further to travel in conditions which may, in an emergency, be
quickly deteriorating.

In relation to normal flight, seat location also determines the
distance from the toilet compartment for those able to utilize this
facility.   This may not always be compatible with a reduced
distance to the exit.

Disabled passengers will, of course, be allocated those seats which
are fitted with movable arm-rests.   However, unless all seats are so
fitted, the problem remains of which seats in the aircraft should be
selected for this modification.

The need of some disabled passengers for greater leg-room has
led some airlines to seat them in the front row where additional
space is available.   Others have considered such allocation as being
in contravention of the need to keep exits clear of obstruction.

When the question of the seat row has been answered satisfactorily, there still remains the question of whether a window seat
or an aisle seat is more suitable. Given that it takes considerably
longer for a physically handicapped or obese passenger to reach an
exit from a window seat because of the time taken to move from
the seat to the aisle, then an aisle seat appears preferable (B13).
However, this does not take account of the likelihood of a disabled
passenger suffering injury from able-bodied passengers seated inboard
pushing past in an emergency in an attempt to escape.

With regard to non-ambulatory handicapped passengers, a study
showed that faster flow rates and speedier total evacuation of the
cabin were achieved when "incapacitated passengers" (anthropological
dummies) were placed at the farthest point from the exit than when
they were close to the exit. When the "incapacitated passengers"
were placed near the exit, six passengers had evacuated within 20s
whereas positioning them farthest from the exit resulted in 17
passengers being clear of the exit within that period (B13). These
results indicate that locating non-ambulatory passengers farthest
from the exit is likely to prevent delay for ambulatory passengers
and allow time for positioning the non-ambulatory passengers for
transport through the cabin. In the light of these results, the
decision to place three chair-bound passengers close to the doors in
preparation for an emergency evacuation suggests a procedure which
may not be most beneficial for the totality of passengers (F11).
This evacuation was completed in 78 seconds. However, there was
advance warning of the evacuation, all eight exits were available
for use, and no smoke or fire impeded the evacuation. In less
benign conditions, the location of the wheelchair passengers might
have caused delay.

There is some confusion and lack of clarity with regard to the
optimal seating of disabled passengers. A clear-cut policy would
allow more appropriate allocation at check-in. If, for some reason,
mistakes occurred at that stage, re-allocation in the cabin could be
carried out more effectively. Such a policy would have to be
firmly grounded in valid research evidence to be accepted. The
militant actions of some blind passengers in the US has focused
attention on the response of some groups to what seem to them to
be unjustifiable rules (N29).

Carry-on baggage

Information regarding the permissible amount of hand baggage
allowed in the cabin is normally made available with passenger
tickets, at check-in points, and elsewhere. Typically, the items
listed comprise one piece not larger than 20in x 15in x 10in (508mm
x 381mm x 254mm), a small handbag or purse, a coat, an umbrella
or walking stick, a pair of crutches, and a small camera. Whilst it
might be interesting to speculate how, for example, a fully-laden

Boeing 737-200 would cope with a full complement of passengers and allowable hand baggage including 130 pairs of crutches, of much greater practical concern is the amount and type of baggage actually brought into the cabin.

The sight of passengers struggling aboard with bulky brief-cases, large garment bags, and newly-acquired souvenirs and tax-free goods is quite commonplace.   Airline staff frequently permit such behaviour without comment, with the inevitable result that the practice continues.   Cabin staff may go to considerable lengths to assist.   One survey found that large or oddly-shaped items were stowed in toilet compartments, on the flight-deck, or on empty seats (C22).

Regulations are based partly on the assumed weight of the loaded cabin, which enters into aircraft performance calculations.   Of greater importance in most instances is the question of the secure stowage of baggage within the cabin, and the avoidance of hindrance to efficient evacuation.   (See Chapter Eight).

Regulations in almost every country are devoted to the need to avoid the dangers arising from loose articles in the passenger cabin. In the US for example, a regulation specifies that passengers shall not be permitted to board the aircraft if their carry-on baggage exceeds that which is specified in the carrier's licence, and that no aircraft may take off or land unless each item of baggage is properly stowed (FAR 121.589).

Since regulations are neither widely understood by passengers nor regularly enforced by airlines, occasional confrontations are inevitable when objections are raised to a passenger's baggage. Obviously, such confrontations are not welcomed by an airline particularly when passengers claim, with justification, that rival airlines raise no objections to comparable items.   In the long-term interest of flight safety and of good relations between airlines and the travelling public, the solution lies in the adoption and regular enforcement of a well-publicized set of regulations regarding the items which may be taken into the cabin.

The survey referred to above recommended that each passenger carry no more than two items, each with a maximum weight of 15lb (7kg) and of a size to be accommodated under the seat or in a designated stowage area, and that aircraft cannot be moved until each item has been properly stowed and the cabin secure (C22).

Smoking

Smoking is prohibited during take-off and landing and whenever the no-smoking sign is illuminated.   Smoking is permitted only for seated passengers.   The prohibition concerning smoking whilst moving about the cabin and in the toilet compartment is communicated both orally by the flight attendant and by an illuminated sign outside the toilet compartment.   Yet it is widely known that passengers do smoke cigarettes in the aisles, and in the toilet compartment. One commentator has reported encountering "on a fairly regular basis the passenger who is seated in the 'No Smoking' section of the aircraft and who goes to the washroom to have a cigarette. The cigarette butt is thrown into the waste bin and we are faced with an in-flight fire" (D10). It should be added that for many smokers the presence of an ash-tray in the compartment may act as an indication that at some stage designers assumed that people might wish to smoke there, and that restrictions were therefore not concerned with safety.

The introduction during 1988 in the United States of a total ban on smoking on domestic flights of less than two hours, followed by a similar ban on all domestic flights in Australia and Canada, may have led to an increase in smoking in the toilet compartment although this is forbidden in any circumstances. The requirement for the installation in this compartment of a smoke-detection device which activates an alarm in the cabin is intended to reduce the danger of fire and to identify the offender. However, some airlines have found that ceiling-fitted smoke alarms have been disabled by passengers wishing to smoke undetected. In view of the difficulties experienced in ensuring compliance with the current rules, it is clear that there is a problem of covert smoking which may well persist in the face of total prohibition. As smoking in the population is increasingly associated with the lower socio-economic groups, it may be that the problem of covert smoking is one most likely to be encountered in aircraft utilized in low-cost charter operations.

INFORMATION AND INSTRUCTION

Certain information must, by regulation, be given explicitly to the passengers both orally and by means of briefing cards.

Contents of oral briefing

At the pre-take-off oral briefing, the passengers are informed of the rules relating to smoking, the use of seat belts, the position of seat backs and tray-tables during take-off and landing, and the stowage of hand-baggage under seats during take-off and landing. They must be informed of the location of the emergency exits, and, where relevant, of life jackets and other devices for use in the event of a ditching. They must be instructed, where relevant, in the use of oxygen masks and the donning of life jackets. This oral instruction is supplemented by briefing cards to which reference is made in the pre-take-off briefing.

After each take-off when the seat belt sign is no longer illuminated, the FAA requires that an announcement is made that passengers are advised to keep their belts fastened when seated even when the sign is turned off. After the no-smoking sign is turned off, passengers are briefed again about the rules relating to smoking.

Those passengers who are not independently mobile are briefed separately, together with their helpers, concerning the routes to the exits and the most appropriate time to begin moving to an exit in the case of emergency.

Before landing, briefing similar to that of the pre-take-off is given relating to seat belts, upright seat positions and stowed tray-tables, smoking, and stowed baggage. After landing, passengers are advised to remain in their seats, with seat belts fastened until the aircraft has come to rest and the engines shut down. This advice should be accompanied by an explanation that any unanticipated stop could cause injury to passengers who are standing. A signal indicating when it is safe to move about is desirable (F01).

Contents of briefing cards
"Each card .... must contain information that is pertinent only to the type and model of airplane used for that flight" (FAR 121.571). This card must illustrate the brace position for all seat orientations in use, show the location of the emergency exits, describe the operation of seat belts and emergency exits, and instruct in the way to don oxygen masks and life jackets.

The effectiveness of safety briefing
The safety of passengers is enhanced if they know what to do in an emergency. Laboratory studies have shown that individuals perform better if they have received instructions about the use of emergency equipment (J10). There is also evidence from accident reports that survivors of accidents are frequently those who have prepared themselves in advance for an emergency. This is likely to be a minority. A survey has shown that less than 10% of passengers looked at the briefing cards while on board (F14).

A questionnaire was distributed to the passengers who had been obliged to evacuate a Boeing 747 after an accident involving fire. Out of 165 passengers, 114 responded to a question concerning the briefing card. Of the 63% who had not read the card, 56% were injured in evacuation-related causes whereas of the 37% who had read the card, only 17% were injured in evacuation-related causes (N07). Whilst caution is always necessary when interpreting such responses to questionnaires, these data suggest a highly significant relation between injury and prior briefing.

Another study showed that after a decompression, only 2 out of 182 passengers were able to put on their oxygen masks and initiate the flow of oxygen. These passengers had read the safety briefing card and listened to the flight attendant's briefing (N22).

When a Boeing 707 crashed in the Pacific, only 5 out of 101 passengers survived. All of these claimed to have read the briefing cards and listened to the safety briefing and all had evacuated the aircraft by means of the overwing exit. Other passengers in their vicinity attempted to exit through the doors and drowned (N12).

PROBLEMS OF BRIEFING

The problems associated with oral briefing and with the briefing cards are, first of all, how to ensure that passengers listen to the oral briefing and read the briefing card. The second problem is how to present the information such that it is readily assimilable and usable in the event of an emergency.

Oral briefing
There have been criticisms of the manner in which the oral briefing is presented. "The whole manner of delivery of emergency briefings tends to play down the significance of the information being presented" (N07). However, it is perhaps understandable that the presentation of the oral briefing does not capture passengers' attention. Flight attendants know that the emergencies which are the subject of the briefing are rare events and therefore unlikely to be encountered. Because of this, their recitation of the standard script may lack conviction. Some passengers may believe that the sophisticated traveller, having heard it all before, will not feel the need to hear it again, and in order to appear well-travelled will attend to other matters during the briefing. Some passengers may be fearful of such an emergency taking place and, by not listening to the briefing, block the unpleasant emotions that might be aroused. Some passengers may believe that briefing information need only be assimilated during an actual emergency situation. The lack of response on the part of the audience is fed back to the flight attendant, reinforcing the bland quality of the presentation.
To counteract the possibility of staleness in the delivery of the oral briefing, taped audio presentations accompanied by demonstrations by flight attendants of the survival equipment may be given. Where video equipment is available, audio-visual demonstrations have been used. While there is no reason to suppose that watching a recording will necessarily be more compelling than watching a live performance, the advantages of such a presentation, apart from the lack of "staleness", are that the use of close-ups can better illustrate the methods of operating emergency equipment which are more difficult to describe in the standard presentation. However, some seats in most contemporary aircraft prohibit an adequate view of the screen.
In an attempt to determine the reasons why some passengers

attended to safety information and some did not, a sample of Californians who had flown at least twice in the previous two years was interviewed (J10). A total of 70% of those participating said they attended to the oral safety briefing and read the briefing card while 30% said they did not. There was agreement in both groups that although the probability of an accident was low, safety information was needed because most of the passengers would survive an accident and could do something for their self-protection in the event of an emergency. Both groups agreed that the safety information presentations were useful because crew members may not be available to assist them following an accident. There was general agreement within both groups that the oral briefing alone was inadequate and that the briefing card was necessary. They also agreed that, following an emergency, there would not be enough time to get instruction from the crew or from the briefing card.

These responses suggest a realistic and responsible attitude towards safety briefing on behalf of the majority. However, some doubt is cast upon this view by the agreement of respondents with the statement that they already knew the safety information on the briefing card before they boarded the aircraft. Considering the variety of aircraft and of configurations, and that the minimum number of recent flights to qualify for inclusion in the sample was two, this was regarded as a rather unrealistic belief on the part of the respondents and one worthy of further investigation.

Those who claimed they did not attend were more likely to be men, younger, better educated, and with more flight experience than those who said they did attend; they were also more likely to fly alone and on business trips. Yet it should be noted that younger men have been shown to have a higher probability of surviving than older men and women (S18).

Two methods had been used to select the sample. In the first, a random selection of potential participants was made from the telephone directory to whom a letter was sent explaining that they would shortly receive a call to interview them concerning aircraft passenger safety. To ensure that everyone with a telephone would have an equal chance of selection, the second method used was to approach respondents by dialling numbers at random.

Those who had received a letter in advance of a telephone interview requesting their cooperation were more likely to claim that they attended to safety briefings and information cards; those who were approached through random telephone dialling were more likely to claim that they did not attend. It is possible that advance preparation may favour the elicitation of responses which are socially desirable.

Very different results emerged from a survey of 62 US commercial flights reported by the Flight Safety Foundation. In this investigation "less than 10% of all passengers looked at their seatback emergency instruction cards while aboard the aircraft" (F14).

Design of briefing cards
The design of the briefing cards presents a challenge not least in accommodating the information and instruction within a relatively small space. Guidelines issued by IATA stressed that briefing cards should be designed to be understood by passengers who are totally unfamiliar with aircraft and safety equipment, and who may have a limited understanding of any of the languages used (I03). For these reasons, descriptive coloured pictures were considered more appropriate than written instructions. Where written text was considered necessary, the guidelines advised that the use of aviation technicalities should be avoided and the number of languages confined to a maximum of three, namely that of the state of registration of the aircraft, English, and one other. Such factors as typeface and colour contrast were considered to be important in relation to the need to be able to read the card under the low levels of illumination likely to prevail in emergency conditions. It was felt that the order in which the information and instruction is presented should reflect the sequence of emergencies to which a passenger might be exposed in the course of a flight.

A study investigating the display of information on briefing cards found that nearly all of a group of 24 undergraduate students who had experience of flying with a commercial airline knew the location of the briefing cards in the aircraft seat pocket. However, they could remember an average of only three information topics out of a possible eight. The most frequently remembered items were oxygen equipment and exit locations, but only as included items, not as remembered information. The students were also required to rank different designs of briefing cards according to the acceptability of their presentational styles. Guidelines for the design of briefing cards based on the results of this study were suggested (A05):

- Pictures with a minimum number of descriptive words are more acceptable than pictures alone, words alone, or pictures with a large number of descriptive words.
- A realistic understandable picture of good quality is preferable to an abstract drawing.
- Where a sequence of actions is called for, two or more numbered pictures are desirable.
- A simple, uncluttered, systematically-organized card format enhances acceptance by the reader.

If the use of words is essential for comprehension of the information, however, this will discriminate against those who, for whatever reason, are unable to understand the words. In an attempt to

accommodate those passengers who are illiterate, who have difficulties in reading text, or who are unable to understand the languages
on the card, the use of text has declined over the last two decades.
However, it should be recognized that pictorial representations are
dependent upon cultural traditions and where briefing cards are to
be used by passengers of non-Western societies, differences in cul-
turally-determined interpretations of pictures must be investigated
(S14).

It is apparent from the examination of briefing cards that there
is a lack of correspondence between some of the pictures and the
reality which would be encountered by passengers in an emergency
situation.   While the intention is, no doubt, to provide greater
clarity, this may be counter-productive if passengers are misled by
the instructions.   The "artistic licence" of the briefing card was
the subject of comment in an accident report (A02).

Some typical examples illustrated in Figures 9.1-9.3 include the

FICTION

FACT

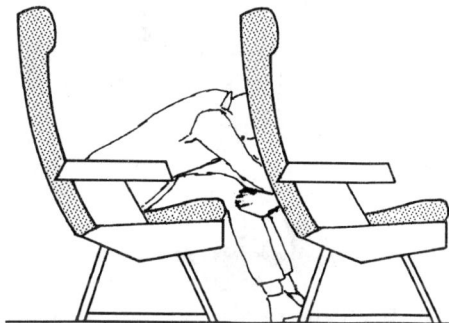

Figure 9.1 Some recommended "brace" postures create problems for
taller passengers.

following.

The brace position is frequently shown by the drawing of a passenger in a seat where the pitch is considerably greater than that in First Class accommodation.   The difficulty, or even impossibility, of assuming this position has been discussed in relation to seats (p. 119).

The passengers donning life jackets are typically shown standing up, with apparently no space restrictions. It is virtually impossible to adopt this posture between blocks of high-density seating, where an aisle area of about $0.3m^2$ ($40in^2$) might serve 6 seats.

Attention was drawn in an accident report to the pictorial briefing relating to a removable overwing exit panel, citing "a large area in which to stand to remove the hatch" and "the hatch being

Figure 9.2 Whilst briefing cards typically illustrate a standing figure donning a life-jacket, it is likely that the task must be carried out in a seated position.

placed on the... seats, with armrests raised" (A02).   Several misleading impressions were given by this illustration, including the
implication that the seat row would be empty, that the arm-rests
would be raised, and that the person carrying out the operation,
wearing what appeared to be uniform, was a member of the cabin
crew.

Figure 9.3 Seats appear to have been vacated and armrests removed
to permit disposal of the emergency escape hatch.

Design of placards
Placards provide both information and instruction about items of
Hardware in the cabin.   Placards indicate such features as the
location of exits and the maximum weight which a locker is
designed to contain.   They also provide information about the
location of emergency equipment and instructions for its use.
There are recommendations concerning the design of placards for
emergency equipment (S22).   These relate to the medium by which
the information is presented, the conspicuity of the placard against
the background, characteristics of lettering and pictures, and the
location of the placard.   Pictures are preferred to words, and words
are to be used only if pictures alone are inadequate.   Placards
should be conspicuous either by means of colour contrast with the
surroundings, by illuminating them, or by enclosing them within a

border.   Upper case lettering is recommended for use in locational placards, and either all upper or upper and lower case lettering on instructional placards.   Details are provided concerning the minimum size of letters and picture elements to ensure their visibility under various levels of illumination, together with the legibility rating of different colour combinations for figure/ground relationships. Recommendations also relate to the location of placards such that they are readily visible and, if two are close together, such that they do not confuse the reader.

Recommendations of this kind can by their nature be no more than general guidelines.   While these are useful in preventing any gross error of information design, there remain many pitfalls in relation to the detailed design of placards where specialist guidance about the content and presentation of information is required.

Pre-emergency oral briefing

Not all the information necessary to ensure safety is provided at the pre-take-off oral briefing.   Where there is time available in an emergency before action must be taken, additional briefing of passengers is carried out.   This concerns the adoption of the brace position, the continued wearing of the seat belt, the loosening of tight clothing, the location of the exits and how to use them, and the importance of leaving behind all hand baggage.   Passengers may be told to change their seating positions.

If there are more exits than flight attendants, able-bodied passengers must be briefed concerning the operation of the emergency exits (the overwing exits are most likely to be opened by passengers), the inflation of the slide (if this is not automatic), and any other responsibilities allotted to them (for example, standing at the bottom of the slide and helping passengers as they land on the ground).

There is some evidence that passengers believe that the appropriate time for attending to safety briefings is when the emergency is about to happen, and that it is a waste of time to attend too far in advance (A05).   This belief is not without foundation.   When written instructions combined with audio-taped instructions were presented at the time of a simulated decompression, all those taking part were able to obtain oxygen (J08).   This compares very favourably with the responses of passengers in genuine decompressions.   Nevertheless, dependence on instruction at the time of an emergency is not an advisable strategy.   In the dark, in smoke, or in the confusion that is associated with emergencies, the likelihood of benefiting from more than the simplest commands of a flight attendant is remote.

Where the emergency is sudden, the flight attendants' communications will take the form of commands.   These are typically terse.   Short, simple commands are readily comprehensible (as long as the language is understood). The use of the negative ("don't..") should be avoided as the initial negative component may not be heard and passengers will perceive the commands as the opposite of what is intended. In order to secure attention and compliance, commands must be given in an authoritative manner, and, if necessary, shouted out. A megaphone may be necessary in some instances.

SOFTWARE AND EMERGENCY BEHAVIOUR

The rules, information, and instruction are intended to safeguard the passengers from hazards; to enable them to avoid behaviour liable to jeopardize safety and to make full use of the Hardware devices provided for their protection; and to facilitate their swift and safe evacuation from the aircraft on to either land or water. Upon the effective utilization of this Software may depend the lives of the individuals on board the aircraft.

Survival in water

Pre-take-off oral briefings for flights over water must include a demonstration of the donning of a life jacket. The briefing card must contain instructions on how to remove flotation devices, if installed, and use them in the water or it must indicate where the life jackets are stowed, how they may be removed and donned, and how to use the inflation systems and survivor locator lights. For extended overwater flights, instructions concerning the use of life rafts (inflation, launching, boarding, and detaching) must be provided. In spite of oral briefing and briefing cards, less than 25% of passengers knew that their life jackets were under the seat (F14).

Some life jackets carry the instructions for donning on the back of the jacket, others carry them on the front but in such a way that they are upside-down to the wearer when the jacket is placed over the head.

Donning life jackets presents many problems. One passenger who survived an unplanned water landing commented "I must say that these things are a helluvalot clumsier to get on than they appear to be in the demonstrations....I had to stop myself and say....there has got to be an intelligent way to get in this thing, without choking myself on the straps" (N19).

An incident involving a Lockheed L-1011 which prepared to ditch after losing power in all three engines, provides further illustration of these problems. This aircraft recovered one engine and was able to make a normal landing. Subsequently, the passengers (of whom nearly 60% responded) were questioned about their experiences when they were preparing to ditch. Of those who responded, 27%

reported difficulty in locating the storage compartment, about 50% said that they were not aware that the life jacket would be folded and sealed in a plastic container, 18% found it difficult to open the container and 68% could not don the jacket while seated with the seat-belt fastened (N21).

Some of these problems have their origin in poor interfacing between the Hardware and the Software, that is between the equipment and the associated instructions. Finding the storage compartment under the seat is more difficult when the seat pitch is small which was the case in an accident in which 17 out 30 occupants died (C09).

An analysis of the task of donning one particular design of life jacket after it had been removed from its storage container and taken from its package indicated that six discrete steps were involved (J06). The first three (pull the jacket over the head, snap the hooks to the D-ring, tighten the straps) are concerned with donning the jacket; the second set of three (pull the inflation tabs, blow into the inflation tubes if the jacket does not inflate, pull the battery tab to illuminate the light on the jacket) are taken after leaving the aircraft.

In order to compare the effects of instructional styles, four groups of people were each given information in different ways:

Group 1 had only the instructions which were printed on the life-jacket itself. This group was regarded as the equivalent of passengers who do not watch, or cannot see, the flight attendant's demonstration.

Group 2 saw a video-tape of a demonstration at an apparent distance from the demonstrator of 30ft (9m), the distance a passenger at the rear of the cabin might be from the flight attendant. This group was regarded as the equivalent of those who attend the safety briefing.

Group 3 saw the video-tape at the same apparent distance as did the second group but each step was also shown in close-up, at the apparent distance of 4ft (1.2m).

Group 4 was shown the start of each of the steps at 30ft (9m); a zoom lens was used to approach to 4ft (1.2m), zooming back again at the completion of each step.

Each group member had the task of unfolding a life jacket, donning it, inflating it and switching on the light as quickly as possible.

In Group 1, only four of the twelve individuals were able to complete the task, and those who did so took an average of nearly three minutes. Half of Group 2 (7 out of 14) were able to complete the task, taking an average of nearly two minutes.

Around half of each of Group 3 and Group 4 were able to complete the task and those who did took an average of less than a minute and a half.

These results demonstrate that different briefing techniques have a significant effect upon performance. However, none of the techniques employed led to satisfactory results, and there is a clear need to use more effective methods, such as "hands-on" training.

The two major errors comprised failing to snap the straps at the front of the jacket and failing to tighten the straps when the hooks were snapped. Both these errors would lead to an unsecured jacket which could easily come off in the water.

Regulations introduced by the FAA require that passengers, after viewing the flight attendant's demonstration, are able to don a life jacket in 15 seconds.

Once donned, life jackets must be inflated to be effective. In general, the instructions regarding inflation are that life jackets should be inflated outside the aircraft. This is because inflated life jackets add considerably to the bulk of the individuals and may restrict egress particularly from the overwing emergency hatch, thus causing unnecessary delay. In addition, there is a risk of puncturing the jacket during evacuation. A further reason for not inflating the jacket inside the aircraft is that the cabin could fill with water so quickly that a passenger in an inflated jacket would float to the ceiling and be stranded there, unable to dive and reach the exit (J13).

There are, however, advantages in inflating the life jacket before leaving the aircraft. If passengers sustain injury during the impact, it might be difficult or impossible for them to pull the tabs which cause the life jacket to inflate. Once in the water, it is difficult to see the tabs, nor are they readily visible in the dark. Airline policies vary in respect of inflation within wide-bodied aircraft.

One essential instruction to passengers who have time to prepare themselves for ditching is that they should re-fasten their seat belts after they have donned their life jackets. It is difficult to don a life jacket wearing a seat belt, so that it is highly probable that passengers will need to be reminded of the necessity for a snugly fastened seat belt. In a ditching of a McDonnell Douglas DC-9, passengers were not briefed to re-fasten their seat belts and the five passengers without belts were thrown out of their seats and one of them was fatally injured (N04).

Survival in water is considerably enhanced if survivors are able to use a life raft. Briefing information regarding rafts is provided on briefing cards but the amount of this information is usually limited to a drawing of a raft with passengers stepping into it from the wing, or in the case of slide/rafts, stepping from the exit into the slide/raft and arranging themselves upon it. Where slide/rafts are in use, then the operation of the exit is probably the most important instruction. Once the passenger is on the slide/raft, it only remains to cut the raft loose from the aircraft, and if this is not done, then it will take place automatically as the aircraft sinks.

By contrast, the use of dedicated rafts found in narrow-bodied aircraft requires the active involvement of the cabin crew to remove the raft from storage, transport it to the exit, attach it to the aircraft, inflate and launch it.

Decompression

If an emergency oxygen supply is fitted in the aircraft, passengers will normally be briefed orally about its use, and information will be given on the briefing card. For flights above 25,000ft such briefing is required by regulation. A survey has shown that less than 15% of passengers fully understood the briefing concerning the use of oxygen (F14). This is confirmed by the evidence that passengers do not respond appropriately to the appearance of oxygen masks. Only a small fraction may don the masks and even fewer activate the oxygen flow. In a decompression in a McDonnell Douglas DC-10, only 2 out of 53 passengers correctly removed their masks from the seat-back compartments to activate the generators and then donned them. The remainder either did not react or they leaned forward and attempted to breathe without fully removing the masks from their stowed positions (N22). In another incident of decompression, only 2 out of a total of 182 passengers donned their masks. Flight attendants had to help the other passengers (N22). The importance of prompt and effective action in the event of a decompression is underlined by the finding that 14 out of 20 individuals taking part in a study failed to recognise that their performance was deteriorating (B28). Flight attendants wearing oxygen masks are unable to speak clearly thus repetition of the oral briefing is not likely to be effective. The extent to which they can assist passengers is limited to giving physical help to those in their immediate vicinity.

When a mask fails to deploy automatically, action must be taken either by the passenger moving to a vacant seat where the mask has deployed, or by the flight attendant deploying the mask manually. Delays may result if passengers are not properly aware of the danger to which they may be exposed and of the consequent need to act swiftly to find an unused, deployed mask when their own is not available.

The effectiveness of placarded instructions on the use of oxygen masks deployed from seat-back compartments was investigated by comparing the performance of two groups, one using the instructions and the other without the instructions (J07). The two groups were alike in their previous experience of flying; they were not briefed about the purpose of the exercise, nor were they given a demonstration of the use of oxygen masks. The results showed that the instructions had no effect on the mask-donning performance and that no more than a quarter of either group were able to don the masks and activate the oxygen flow successfully.

Subsequent development and testing of instructions which were attached to the inside of the oxygen mask-compartment door showed that 94% of those taking part were able to don the masks and activate the flow of oxygen, though the average time for this was 40s with a range from 12 to 122 seconds.

While this is a significant improvement on the performance of uninstructed individuals, the time taken must be considered in relation to the time of useful consciousness which averages 54 seconds at 34,000ft for inactive individuals (B27). By contrast, highly trained individuals are unlikely to require more than 12 seconds to don oxygen masks even when faced with design problems (B28).

Where oxygen masks are stored in the roof panels of the aircraft, or on occasions when there is insufficient illumination to read instructions, an alternative method of presenting instructions is required. Recorded instructions transmitted through the public address system have been found to produce results similar to those obtained with instructions displayed in the mask compartment (J08).

It may be concluded from these studies that instructions provided at the time of the decompression are more effective than the pre-take-off oral briefing and the instructions on the briefing card. However, such instructions may not be heard by passengers. The noise accompanying a decompression may be sufficiently loud to make the public address system inaudible and the fogging of the atmosphere may make reading difficult.

The type of errors made in attempting to use oxygen masks will indicate the areas that the instructions should address. To be effective, the mask must cover the mouth and the nose. There is a tendency to cover only the mouth. In most aircraft, the oxygen flow has to be activated by pulling the mask to the face. There seems to be a reluctance to pull the mask and, instead, passengers put their faces up to the mask thereby failing to obtain oxygen. Instead of adjusting the strap around the head, the mask is held against the face. This is inadvisable. As oxygen-impoverished blood may take up to fifteen seconds to reach the brain from the lungs, the individual may become unconscious before being able to benefit from the use of the oxygen mask and, as a consequence, drop the mask, becoming vulnerable to further effects of hypoxia. An important consideration for those in charge of young children or elderly persons is the requirement, contrary to a caring tendency, to attend to themselves before attending to others. Cabin staff must do the same.

Emergency evacuation

The oral briefing provides information on the location of the emergency exits but not on their method of operation. On the briefing card, the exits are shown together with their method of

operation. Experience has shown that passengers do not memorize
or apply this information in an emergency.

In a comparative study of three aircraft crashes in which all the
deaths were caused by fire or smoke, it was found that those who
survived the accident sat, on average, closer to potentially usable
exits than did those who died (S18). The authors suggest that seat
location may be an advantage in "mild" fires where the seat
location determined the position of the passenger in the queue
formed to evacuate.   However, among both survivors and fatalities,
many passengers tended to sacrifice some of their initial locational
advantage by ignoring nearby exits in favour of more distant ones.
The reasons for this, as adduced from survivors, were in part
rational in that their nearest exit was blocked by fire, hidden by
smoke, or unopened at the critical time.   Other reasons included
the felt need to "join the crowd" heading for a more distant exit, a
lack of familiarity with the aircraft and thus ignorance of the
proximity of exits, and a tendency to attempt to go out by the
same door through which they boarded the aircraft.   The authors
commented on the "rarity of instances in which a passenger, once
impelled toward a given exit, changes his mind and chooses
another". It was also apparent that the survivors had made more
effective use of available exits than had the fatalities (S18).

In addition to the instructions on the briefing card, the methods
of opening the exits are described on (or beside) each exit.

The Type III exits (see Chapter Eight) are provided in order that
emergency evacuations can be carried out at an acceptable speed.
In view of the inconvenience in their use, and the increased risk of
injury, some operators elect to avoid their use during a
"precautionary" evacuation.   Flight attendants are not usually
stationed beside them and as a consequence Type III exits are most
likely to be opened by passengers. The briefing card for one aircraft
with an overwing exit gave no indication of the weight of the
hatch, showing a female figure effortlessly removing it from the
aperture (A02).

The floor-level exits are manned by flight attendants who are
expected to operate them after ascertaining that conditions outside
are suitable for evacuation.   It may be necessary, however, for
passengers to open these exits if flight attendants are injured, or if
they are not at their stations.   There is a possible danger that
passengers may inadvertantly disarm the slide in their attempts to
open the exit which could make the exit unusable.

While the extent of the variation in exit operation both within
and between aircraft underlines the need for clear instructions and
for adequate illumination, there is a greater need for
standardization in the operation of equipment of any kind that is to
be used in an emergency.   Standardisation would have the effect of
simplifying crew training and passenger briefings by avoiding

confusion, particularly since placarded instructions are unreadable in the dark or in smoke. However, the introduction of standardisation would be unlikely to be retro-active in its effects. The necessity for good instructions would remain for those occasions when the exit must be operated by a passenger.

Smoke

While smoke hoods may be effective in combating the hazards of smoke and fumes, they will have little value if they are not used properly. This is particularly important in relation to the use of smoke hoods for emergency evacuation in fire where time intervals are critical.

An investigation into the success of wearers in donning the relatively simple Type S smoke hood was carried out in which it was reported that 90% of these individuals encountered some sort of problem, in spite of regarding the briefing as clear (S17). Nevertheless, "all were able to get their hoods on safely and quickly". In response to a signal, they removed the smoke hood from its packet located on the back of the seat in front and put it on. Those who donned their hoods without any problems took an average of 12 seconds while those who encountered problems took an average of 17 seconds. The activities involved in donning the smoke hood comprised removing it from its packet, finding the neck seal, spreading the neck seal, inflating the hood by shaking it out to fill it with air, pulling the hood over the head, getting it over hair, spectacles, nose and chin. Most of the problems concerned inflating the hood prior to donning it so that it would contain a maximum amount of air. Less than half (40%) were successful in fully inflating the hood and 18% had minimally inflated hoods. Other problems encountered by the wearers concerned the difficulties in finding and spreading the neck seal.

Variations in the briefing instructions showed that more positive feelings towards the hood were associated with more information about it. Just under a third of wearers had negative feelings about the hood and the reasons centred on a fear of shortage of air in the hoods. They nevertheless "willingly" donned and wore the hoods. Nearly all said that they would use the hoods in an emergency if requested to do so. It should be noted that experimental subjects exhibit a tendency to comply with the demands of the experimenter and therefore the low level of reported negative affect should be treated with caution.

One interesting finding was that when those who had had problems were given a second opportunity to don the hood they reduced their donning time significantly from just over 17 seconds to just over 13 seconds. The number experiencing problems also decreased from 13 to 8 and the number not fully inflating their hood decreased from 10 to 4. It is interesting to compare these times with the performance standard specified by the CAA (C14) which required that a smoke hood be capable of being put into effective use within 10 seconds of its need being recognised. An uncontrolled investigation to determine the time taken to don a more complex smoke hood showed that, after reading the instructions, the average time taken to open the box containing the smoke hood and then don it was 27 seconds (B29).

The evidence from various studies indicates that it is the smoke and not the smoke hoods which induces panic, and that the effects of the smoke hood were likely to prevent panic because of the absence of respiratory distress (M07).

The effects of smoke hoods on behaviour in the laboratory may not be the same as the effects in an emergency evacuation. For example, experimental evidence about the ease of donning oxygen masks (H08) was contradicted by evidence of difficulties encountered by passengers in genuine emergency situations (J13).

Nevertheless, it is possible to make some tentative statements about smoke hoods and behaviour, based on the experience of passengers with other pieces of safety equipment. It is evident that passengers have problems in coping with these items. Difficulties have been reported in the location and donning of life jackets, in the use of oxygen masks, in the operation of seat belt buckles, and in the opening of emergency exits. It would seem likely, therefore, that passengers would experience problems in the location and donning of smoke hoods. These problems are not, however, insoluble.

The timing factor makes smoke hoods more similar to oxygen masks than to life jackets. It is advisable in the event of ditching to take a life jacket even if it is not donned in advance of leaving the seat, as it will certainly confer advantages in the water. An oxygen mask, however, must be used as soon as it is deployed to counter the effects of hypoxia. Similarly, a smoke hood is only useful inside the cabin and must be donned in advance of the entry of toxic smoke and fumes. This raises problems if the simplest form of smoke hood is used, as it is not likely to be effective for more than two minutes. If a more complex smoke hood is used which contains an air supply and thus allows more time for evacuation or for use on board during an in-flight fire, then there are questions of standards to be determined both in relation to the design of the hood and the instructions for its use.

In general, safeguards which do not require voluntary action on the part of individuals are more effective than those which do. The improved fire-hardening of materials used in the construction of the cabin and the development of a sprinkler system in the cabin to combat the effects of fire are together likely to be a much more satisfactory solution to the problem than the provision of smoke hoods (B11). The reactions of passengers, to the extent that they may be identified from a survey of 30,000 air travellers from 100 different countries, tend to favour this solution in placing fire-resistant cabin materials as their highest priority in safety, followed by stricter controls on hand baggage and better access to emergency exits. Smoke hoods were sixth in their list of priorities (G20).

SOFTWARE FOR CABIN CREWS

Standard operating procedures

A great deal of the behaviour of cabin attendants follows pre-planned procedures, or "drills", which are learned during the training period. Such drills are recorded in the form of lists of actions to be carried out in particular circumstances. All such lists will be included within the manuals supplied to crew members, and carried on board the aircraft.

There are obvious advantages arising from this highly-structured, procedural approach, particularly in the case of reacting to emergencies. Optimal solutions to problems are best achieved in conditions removed from the stresses of urgency and danger. Formalized procedures may be developed after consideration by many different experts who can contribute various viewpoints before a precise procedure eventually evolves. Drills are, of course, always subject to change in the light of additional experience.

A second consideration arises from the absence of stability in the constitution of a crew, the members of which are brought together by a rostering procedure. They may never have met prior to a particular flight. In such circumstances, an effective co-ordinated effort can be achieved only by way of the employment of standardised procedures which are well-known to each individual attendant.

The crew manual will contain lists of actions to be performed in sequence in the event of such emergencies as fires, sudden decompressions, or various types of emergency landings. Each cabin attendant will have been allocated a particular role, including which exit is to be manned, such that emergency procedures can be followed with the minimum of delay. Either the senior cabin crew member or the aircraft captain will provide the signal to initiate particular actions, according to the circumstances. In the case of pre-meditated emergency routines, a particular form of words may be used by the captain over the public address system in order to alert the cabin crew to begin their routines.

A question arises concerning the relative merits of carrying out emergency procedures either from memory or by reference to printed documents. Speed is the obvious advantage of the former method, but the penalty is a lower level of reliability. A compromise widely used is to employ a certain number of memory items to include the most urgent actions, followed by the use of a printed list. When using this method, the first post-memory action should comprise a review to ensure that all the memory items have in fact been fully and correctly carried out.

Whilst the advantages of the use of drills are beyond challenge, a very real question arises concerning the circumstances under which the pre-planned procedures should be abandoned and alternative, adaptive tactics employed to deal with the particular prevailing conditions. It is impossible to prepare for all possible eventualities. Equipment may be damaged and inoperable; key personnel may be incapacitated. Each individual cabin attendant must decide how best to react in the existing situation. In exercising personal judgement, crew members should remain aware of the relevant prescribed procedures and should have clear reasons for deciding to adopt alternative tactics. Decisions made swiftly may or may not appear to be wise in the light of future analysis. However, just as examples may be quoted to illustrate the success of orderly and co-ordinated exercises performed "according to the book", so other examples illustrate the benefits deriving from prompt, creative, flexible decision-taking attuned to particular unexpected circumstances.

Communication

Communications between flight-deck crew, cabin crew, and passengers play a vital role in the performance of those procedures concerned with emergencies. The question of the initiation of an emergency evacuation serves as an example.

One airline's manual stated "Do not initiate an evacuation until the captain says 'evacuate'....If there is obvious structural damage, heavy smoke, or flames, or other crew members have started an evacuation, do not wait for the captain's signal" (N19). Whether or not to signal an evacuation is thus a matter of fine judgement.

Other operators' procedures for premeditated emergency landings require cabin crew to begin the evacuation immediately the aircraft comes to rest, having previously completed a number of preparatory actions.

There have been complaints when flight attendants have waited "excessively long" for the order from the flight-deck (N19). In one accident, when no order was given from the flight-deck after the aircraft had landed, all the occupants died of the effects of an in-flight fire (P05). In another accident, flight attendants waited for a signal from the flight-deck to initiate an evacuation but communication was impossible as the flight-deck had been separated from the cabin (N19).

The question of initiating an evacuation also arises when there is no opportunity for premeditation of the emergency. There are instances of flight attendants initiating evacuation without informing the flight-deck crew whose first intimation of the event came from the illumination of the "doors open" signal (N07). On one occasion, an off-duty flight attendant initiated an evacuation in the mistaken belief that the aircraft was on fire (P03).

All emergency evacuations are potentially hazardous and injuries of various degrees of severity are often incurred. Evacuations initiated by flight attendants will be highly dangerous when engines are running and the aircraft is on the move. Such was the case when a flight attendant saw smoke emerging from an engine and contacted the flight-deck for permission to evacuate. Because the flight-deck crew were preoccupied with the preparations for take-off, there was a slight delay before the flight attendant's call was answered. The form of the response was "Go ahead!" meaning that the flight attendant should start speaking. However, this was interpreted by her as agreement to initiate evacuation, and many passengers were injured as they left the aircraft while it was moving along the taxi-way with all engines running.

This problem has no simple solution. In many situations, the flight-deck crew are intensely preoccupied with the task of handling the aircraft and have no spare capacity for communicating with the cabin. Also, without mirrors or external video cameras, a strictly limited view of the exterior of the aircraft is available from the flight-deck. The captain is therefore dependent on others, such as air traffic controllers or the cabin crew, to provide the relevant information and may therefore not be the best judge of the most appropriate time to evacuate. These factors, together with the need for speed in relation to an emergency evacuation where even seconds are critical, lead to a dilemma.

Passengers have themselves initiated an emergency evacuation. While a Lockheed L-1011 was taxi-ing in preparation for take-off, passengers saw smoke and flames in the engines. This was not an emergency but the effect of residual fuel lying in the engine tailpipe. The passengers at once initiated an evacuation, opening emergency exits and evacuating down the slides while the aircraft was moving along the taxi-way. Many injuries were sustained as a consequence of this action.

A further example of communication breakdown is provided by an incident which ended, fortunately, quite safely (N21). A Lockheed L-1011 flying over water in the West Indies experienced problems with an engine and the cabin crew were informed that the aircraft was returning to its point of departure. About a quarter of an hour later, further engine problems were experienced and ditching seemed inevitable. The senior flight attendant was instructed from the flight-deck to prepare the cabin for ditching. Five minutes after this, there was a message from the flight-deck that ditching was imminent. In view of the height of the aircraft and its rate of descent, more than 6 minutes would, in fact, have been available prior to impact with the water. Fifteen minutes later, however, the flight attendants were told to prepare for a normal landing.

Following the initial instruction, the cabin crew had prepared the passengers for ditching; supervised the donning of life jackets, located able-bodied individuals at the doors, briefed them about the operation of exits, and finally instructed the passengers how to adopt the brace position. On being advised that ditching was imminent, the crew issued the order to brace. This position was maintained during the ten minutes prior to the announcement that a normal landing was about to take place.

The cabin crew were disadvantaged in not being advised of the expected period of time available in which to carry out their preparations. Consequently, some passengers considered that flight attendants had neglected to attend to their needs or to provide adequate information concerning the anticipated emergency. The instruction to brace was given much earlier than necessary, and was not rescinded as soon as it became clear that the emergency had been averted. Recommendations from the NTSB to the FAA followed from an analysis of this incident. These stated that flight attendants should be informed of the approximate time available for cabin preparation in the event of in-flight emergencies, and that there should be a signal from the flight-deck to the cabin indicating when to adopt the brace position.

It may be necessary in some emergencies for cabin crew to request the help of able-bodied passengers. Passengers have on occasion refused to assist. When faced with passengers who are reluctant to cooperate, the flight attendant must rely on persuasive and assertive communication to ensure that the cabin is suitably prepared.

First Aid and emergency procedures

FAA regulations require that "crew member emergency training must provide for the proper use of first aid equipment and the handling of emergencies involving illness, injury, or other abnormal situations involving passengers or crew members" (FAR 121.309).

There is a wide variation in the number of hours devoted to First Aid training. A survey in 1980 demonstrated a variety between different US airlines ranging from 4.5 hours to 28 hours in initial training, and from 0.5 to 8 hours in each session of recurrent training. The time given to training in the particular skill of cardiopulmonary resuscitation (CPR) varied between 0 and 15 hours in initial training and 0 and 4 hours in recurrent training (C22).

Two groups of medical problems will be encountered in flight. The first consists of injuries and illness which have been incurred on board. Examples include food poisoning from a contaminated airline meal, injury from turbulence, burns and scalds from hot items in the galley, and choking on pieces of food. The second group consists of pre-existing conditions, such as asthma, diabetes, or cardiac problems, which may become acute during flight and require attention.

The variation in training periods illustrates the wide range of policies adopted by airlines concerning the level of expertise required of cabin staff. At one extreme, an airline may aim to provide only a rudimentary training. At the other, cabin attendants are trained to deal with a wide range of medical problems. It has been suggested in some quarters that at least one member of each crew should be capable of providing "paramedical" services.

On an aircraft carrying several hundred passengers, it is highly likely that at least one physician will be present on board. It is widespread practice to seek the assistance of such travelling qualified people should this prove necessary in flight. Obviously there are potential difficulties associated with establishing the credentials of personnel volunteering their services, and some risk in respect of legal liabilities. In practice, the occasional call for medical help in an emergency appears to function successfully.

At least one airline has pioneered a scheme which utilizes travelling physicians in a more systematic way. In return for concessionary fares, physicians within the scheme make themselves known to the crew when boarding the aircraft, and remain ready to provide help on request (T05). In this way, the doctors have an opportunity to become completely familiar with the equipment and supplies available on board, and are able to develop some experience in treating emergencies within the somewhat unusual setting of an aircraft cabin.

REVIEW

Each item of Software must interface smoothly with many other components of the system of which it is a part (see Figure 6.1, p.84). Within the organization of a single airline, there exists the legal necessity that its rules and procedures should conform with the statutory instruments of the state of registration. Should the state be a signatory of ICAO, there exist expectations of conformity with internationally agreed rules. Advantages will acrue from further compatibility between airlines such that arrangements for wet leasing or for common training schedules might be facilitated.

Mismatches may arise at S-H interfaces. The sequence in which the items forming such interfaces are selected as system components is variable, with the consequence that the direction of the interface design task is sometimes S to H and sometimes H to S. The skills involved can be quite different in each case. A life jacket, for example, will be designed prior to the instructions for its use. Such Hardware items will normally have a long design history based upon operational experience together with programmes of research and development oriented towards an optimal design solution. The task of the Software writer is to construct a set of instructions for the user, ensuring an accurate and comprehensive correspondence with the features of the Hardware item.

The provision of emergency exits serves as an example of S-H interfacing in the reverse direction. Here, the starting point is a Software regulation which defines the maximum duration of a demonstrated evacuation procedure. The number, size, and location of exits are governed by the requirement that the Hardware is adequate to satisfy the rule.

Both the examples above serve to illustrate a further property of SHEL systems, in that the design of single interfaces must necessarily be part of an iterative process which involves all other areas of the system. An attempt is made in Figure 6.1 to represent this system characteristic of highly interactive components. An ideally-designed system is one in which all the component parts exist and function in harmony with one another. A disturbance at any point may lead to repercussions throughout the system. Consideration of the S-H interfaces involving the examples of life jackets and emergency exits is thus an abstraction from the wider problems of total system design, since Liveware is clearly of the highest relevance in each case. Both the jackets and the exits must take account of the size and shape of Liveware, and the related rules and procedures must interface with human behavioural characteristics. In this context a word of caution is in order concerning the anthropocentric diagram shown in Figure 6.4. Whilst this provides a convenient schematic representation of the four interfaces involving the L components, and appropriately places L at the centre of attention, it fails to illustrate the complex interactions which are an essential feature of systems. Only by coming to terms with the full range of SHEL problems can HF make a satisfactory contribution.

Whilst Hardware expertise is plentiful in industry, far fewer people and much less experience is available in the preparation of Software and in the closely-related field of documentation. The notorious inadequacy of computer documentation illustrates this point. Standardization of Software is generally more difficult to achieve than that of Hardware, and many intractable difficulties lie in the area of Software quality testing.

Whilst in most respects appropriate rules and procedures pertaining to the aircraft cabin have been established, their implementation is far from complete. Violation by passengers of rules relating to Dangerous Goods, hand-baggage, smoking, and intoxication are quite commonplace. Evidence from emergencies indicates that safety briefings are frequently ignored or the information is forgotten. Satisfactorily designed Hardware is sometimes rendered useless by virtue of short-comings in the associated Software.

Recommendations for improving the interfaces between Software and other components are best considered when Liveware aspects are also taken into account. Further discussion therefore appears in the following chapter.

10   Liveware in Emergencies

> The effects of high stress levels upon human behaviour have
> been studied and documented. Cabin staff should have some
> understanding of the varieties of behaviour so elicited in
> order better to manage any emergency which arises.
> Unnecessary death and injury are caused by passengers' lack of
> preparedness. A recognized syllabus for passengers provides
> the only solution.

RESPONSE TO STRESS

An understanding of human reactions to stress forms a central
ingredient of the planning procedures which need to be completed in
readiness for any emergency situations which may arise. Only by
anticipating such reactions does it become possible to exert the
degree of control necessary to minimise undesirable effects.

One obvious anticipatory step involves ensuring that personnel who
might become directly involved in managing an emergency situation
are well prepared for their difficult, and perhaps traumatic, task.
An additional problem here derives from the fact that such
personnel are themselves likely to be subject to stress at just the
time they are attempting to deal with it in other people.

The best results may only be expected of cabin crews if careful
attention has been paid to the procedures of personnel selection, of
training in stress management, and in the regular rehearsal of
strategies under a variety of possible emergency scenarios.

The concept of stress
An obvious difficulty which places limitations upon the study of
stress is the impossibility of conducting controlled experimental
studies with the sole purpose of collecting evidence. Except for
the manipulation of very mild levels of stress by harnessing the
spirit of competition between volunteer subjects, or by offering

rewards for certain levels of achievement, experimental studies are unethical and unacceptable.   Evidence is available, however, from observers of natural disasters and other situations in which people are subjected to various degrees of pressure.   Included in such evidence are eye-witness reports of behaviour in aircraft accidents and incidents.   All such data assist in contributing to our understanding of the nature of stress.

Ordinary language is rich with words describing our reactions to objects or circumstances which give rise to stress.   These stressors may be of a relatively permanent nature or may occur only as sudden transients.   Our reactions may be shared by most of our fellow men or may be highly idiosyncratic.   The stress-induced behaviour may be regarded by ourselves and others as entirely appropriate in the circumstances, or thoroughly irrational and misplaced.

General Adaptation Syndrome

The human species is not alone in exhibiting certain characteristic response patterns in the face of stressful situations.   Much research has been carried out by observing the effects of the presence of stress upon animals, and these studies have given rise to a widely accepted descriptive model termed General Adaptation Syndrome (GAS)(S11).   Physiological research upon the mechanisms involved and studies of the corresponding behaviour patterns of animals have identified three stages of the syndrome.   The first of these is essentially an alerting response arising from adrenal gland secretions. This primary response readies the organism to mobilize the energy which may be required to fight or to flee. In the event that the stressor is quickly withdrawn, normal equilibrium of the body will soon be re-established. If, however, the threat persists then a further, but different, adrenal secretion occurs due to the release of a pituitary hormone.   In this second stage many animal species exhibit characteristic postures, gait, and behaviours which prepare them to react to the perceived threat.   These reactions carry a price tag;   resources are called upon at a higher rate than that at which they can be replaced.   In the event that the stressors persist over a long period of time the animal will eventually move into the third stage of the GAS and become exhausted.   In an extreme case the organism will eventually die.

In the case of the human GAS, the physical and behavioural aspects are enriched by the subjective feelings which we recognise in ourselves, or have heard described by others.   Antipathy is signified by labelling stressors as nasty, unsavoury, odious, offensive, disgusting, revolting, obnoxious, vile or loathsome.   An element of our tendency to withdraw from them is contained in such words as repulsive or repellent.   A further suggestion of fear is contained in terrible, dreadful, horrible, or   ghastly.   With an element of uncertainty about the stressor, we may employ such descriptions as nervous, jumpy, edgy, tense, tremulous, fidgety, or apprehensive. After a level of belief about the existence of a threat has been established we become troubled, anxious, disturbed, disquieted, concerned, worried or apprehensive. As the proximity of the stressor increases we may be alarmed, afraid, scared, frightened, horrified, fear-stricken, or terrified. In these circumstances human beings may exhibit aggression or flight, either of which may be an entirely rational response. An alternative, maladaptive response of inaction is recognized in such words as paralyzed, scared stiff, petrified or frozen.

Although it is possible to describe some general features of the syndrome, people do, of course, exhibit significant differences in their stress behaviour. To some extent these correlate with measurable dimensions of personality, although a single individual is likely to exhibit variability from time to time. The variety of the sources of stress is vast. Such, for example, are the reactions to highly specific stressors shown by some people that no fewer than about seventy different phobias have been named.

Theoretical models
Many theoretical models have been developed to assist in the processes of explaining, predicting, and controlling human behaviour. One of these, the Yerkes-Dodson Law, owes its origin

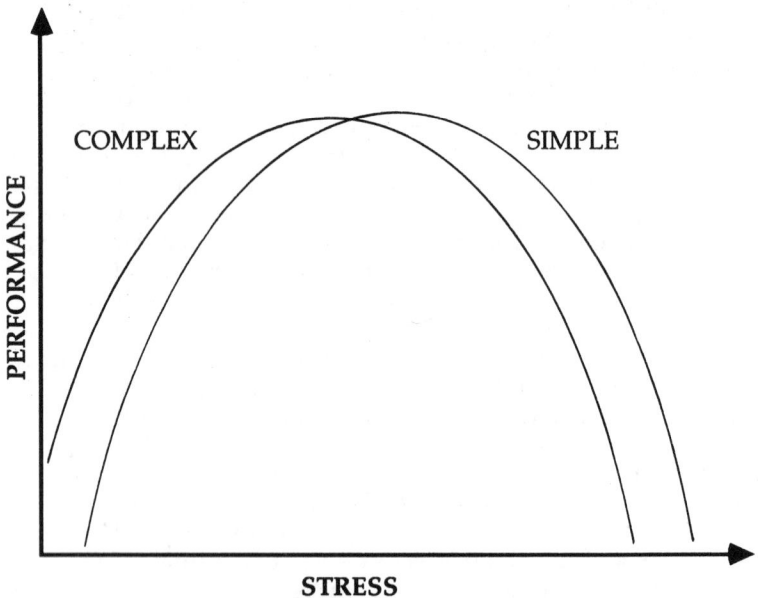

Figure 10.1 The Yerkes-Dodson Law, indicating the level of stress required to produce optimal performance.

to animal studies carried out during the first decade of the century (Y03). The investigators studied the performance of mice during a two-way discrimination task which involved "punishment" when the incorrect selection was made. Analogous studies have since been performed using different animal species and also using human subjects.

The basic model deriving from this work comprises an inverted U-shaped curve relating performance with the reward-or-punishment continuum, which has been interpreted as a motivational scale or level of arousal. At a low level of arousal produced, for example, by a very mild punishment for error, the level of performance was also low, and these variables increased together until some optimal point was reached. Thereafter, further increases in arousal, or stress, brought about by more severe punishment, caused the performance to deteriorate. It was also found that as the difficulty or complexity of the task increases, so the optimal performance level is achieved at a lower level of stress. This model is illustrated in Figure 10.1.

In practical terms, the conclusion to be drawn is that performance reaches its highest levels not in the absence of stress, but given the right level of stress. The person whose arousal level is too low may be complacent or too relaxed. On the other hand, too much pressure will equally cause confusion, panic, or other counter-productive behaviour patterns.

Other, more complex, models such as Catastrophe Theory based upon the topology of the cusp have been advanced as ways of understanding the apparently sudden behavioural changes which can occur under certain conditions (W09). Psychologists make use of such models in their attempts to understand behaviour, and to convey to others a variety of insights which have some practical usefulness within a repertoire of occupational skills.

BEHAVIOUR IN EMERGENCIES

No single pattern of behaviour will necessarily be elicited in the face of an emergency. The variety of human reactions will depend upon the prevailing circumstances, in addition to the temperamental dispositions of the individual people involved. Some people may panic, some may become frozen into immobility, some may seem helpless and dependent. Others may respond in a competent and orderly way. Cabin crews require an understanding of such behaviour in order to be prepared to exert the maximum amount of control necessary in the interests of safety.

Panic
The word "panic" is derived from the Greek deity Pan who was

believed to be responsible for the mysterious noises of the night. Several legends describe the ways in which Pan's activities led to terror and chaos in the ranks of those who heard the misleading sounds.

As the result of smoke in the cabin of a Lockheed L-1011, the flight engineer who had investigated the situation reported that everyone in the rear of the aircraft was panicking. After an actual fire in the cabin had been reported, a cabin attendant informed the captain that the aisles were blocked by fighting passengers. None of the 301 passengers on board survived after the aircraft landed and investigation revealed that the door handles had not been operated (P05).

Two features of behaviour, appearing singly or together, lead observers to the use of the description "panic". The first is that it is hyperactive and often unproductive. People have clawed at metal doors or stone walls in their attempts to escape. Secondly, the behaviour may involve gross violations of the individual's normal standard of social behaviour. Fit young passengers have trampled on weaker people in their efforts to avoid danger.

For these reasons, panic arouses negative moral reactions. People displaying panic are considered culpable and inadequate. Because their behaviour may threaten the survival of the group, they are also considered anti-social. In contrast, moral approbation is attached to behaviour which is calm, resourceful and unselfish.

Persons whose behaviour is described as "panic" are likely to experience intense feelings of life-threatening fear, associated with an urgent need to flee from the source of the fear. In these circumstances, previously learned behaviour styles, both of a social or a technical nature, are inhibited. Quite outrageous acts against other people may be committed whilst at the same time there is an inability to perform such elementary tasks as operating door handles. The disorganization and ineffectiveness of the panicking individual's behaviour is the consequence of high levels of arousal. As the complexity of the tasks increases, the likelihood of effective performance decreases.

There are certain features of emergency situations which appear to trigger the onset of panic. One such feature is the loss of vision, which may be brought about either by a failure in the lighting system or by smoke. There is a widespread fear of darkness, unparalleled by the loss of other channels of sensory input.

Panic has been described as "infectious". In a crowd, some who might otherwise remain relatively calm will, in the presence of those displaying panic, swiftly imitate the response of their neighbours.

Another feature which appears to be important in the genesis of panic is the perception that the escape route is becoming progressively less accessible. In the cabin, such a situation may arise as more people are seen to be blocking the evacuation route.

Behavioural inaction

Media interviewers tend to enquire of those who have escaped disaster, "Was there any panic?" to which the answer is most frequently, "No, everyone remained calm". Indeed, it seems that panic occurs rather infrequently and that behavioural inaction is more likely to be encountered in disaster situations, a result confirmed by evidence from aircraft accidents (J13).

Behavioural inaction or "freezing" occurs when individuals become immobilized as a consequence of extreme terror. The word "petrified" which is often used in this context literally means "turned into stone" and thus unable to think, to feel, or to move. This response is biologically adaptive in some circumstances, for instance when an animal remains quite still to avoid the attentions of a predator. It may be adaptive for the individual to the extent that it operates as a defence mechanism, protecting the individual from the noxious experience of overwhelming fear. It is not adaptive, however, in a situation such as an aircraft accident where survival is more likely to depend upon speedy action. Freezing is likely to result when individuals under severe stress are called upon to carry out a task which is unfamiliar and novel (J14). The inability to perform the required behaviour will increase the stress and in this situation the individual, unable to cope effectively with the demands of the environment, reacts by freezing.

There is some confusion in regard to the phrase "negative panic" which is sometimes used to describe freezing and sometimes refers to behaviour characterised by a lack of urgency in situations where urgency might be expected. Passengers stop to gather their hand baggage and tax-free goods and behave in a casual manner at odds with the reality of the situation. In contrast to freezing, this behaviour is contingent upon a low level of fear in which the full extent of the threat has not been comprehended. Because aircraft emergencies are rare events, there is a tendency to assume that "it can't happen to me" until faced with incontrovertible evidence that it can, by which time valuable minutes have been wasted.

Dependency and helplessness

The unfamiliar circumstances of an emergency, characterized by chaos and the presence of threat, are likely to evoke feelings of considerable uncertainty and anxiety. The individual is motivated to reduce the anxiety, and may attempt to do so by seeking for stability and certainty in the environment. In this situation, familiar behaviour patterns, even though possibly maladaptive, will provide some element of stability. Thus passengers have sometimes retraced the path to the known exit through which they entered the aircraft, regardless of the utility of this action.

Passenger dependency was illustrated in an experimental study designed to test the effectiveness of sequential flashing lights in directing passengers to exits in the presence of smoke. When those taking part were instructed to follow the lights to the nearest exit, 93% did so. However, when this exit was found to be inoperable, six out of seven participants stayed beside the exit though they had been instructed to try to escape. Another group of seven were instructed that, if they found the exit to be inoperable after trying to open it, they should look for another exit. All these participants complied (J09).

In the search for certainty, the individual is highly suggestible and will respond to others who appear to know what to do. This may also lead to maladaptive solutions; aircraft passengers have moved past usable exits to follow other passengers to an exit which is unusable. However, this suggestibility is functional where leaders, formal or informal, who know what should be done to maximize safety, are able to exercise their leadership role.

Competence

Ideally, behaviour in an emergency embodies the characteristics of any skilled response; it is smooth, coordinated, and effective, requiring minimal effort for its successful execution. A high level of skill is consequent upon the appropriate training of selected personnel with adequate rehearsal to maintain a desired level of performance.

THE MANAGEMENT OF EMERGENCIES

The aim of emergency management is to minimise damage to life and limb and to property. The primary Liveware component of any such programme must be concerned to avoid the manifestation of disorganized and ineffective behaviour and of behavioural inaction, and to encourage the performance of activities which will ensure survival of the maximum numbers. In the light of what is known about human behaviour and the conditions which favour the development of panic, how can we facilitate the emergence of adaptive behaviour? While a detailed programme of action is outside the scope of this book, some general principles may nevertheless be outlined as a guide to the development of appropriate strategies.

Levels of fear and stress

The different potential hazards encountered in flight lead to different behavioural requirements to safeguard passengers' well-being. The risk of turbulence, for example, demands only that passengers should remain in their seats with their seat belts securely fastened.   The survivable impact, on the other hand, requires far more positive responses in a situation where the timing of various actions may be of critical importance; the cabin must be evacuated speedily as there is the ever-present possibility of fire.

In the face of danger, in the cabin or elsewhere, there is an optimal level of arousal, stress, or fear at which behaviour is integrated and purposeful.   It is evident from Figure 10.1 that panic behaviour and freezing are the consequences of high levels of stress.   In the case of panic, behaviour is ineffective and disorganised; in the case of freezing, the individual apparently ceases to function. One task of a group leader, such as a cabin attendant in an aircraft emergency, is to foster the appropriate level of stress having regard to the possibility of the need for action.   In practice, this means that extreme values of stress are best avoided as the resulting behaviour is unlikely to be adaptive. The appropriate level which will best facilitate the emergence of adaptive behaviour is a matter of judgement, developed from experience.   One method of influencing levels of stress is by means of information.

Information

Some emergency situations, such as explosions on board or impact with high ground, take place suddenly without warning.   Others, such as an in-flight fire or impending ditching, develop over a longer period during which there may be some uncertainty about the existence of the threat, or of its magnitude.   Passengers may then suspect that something is amiss before the arrival of reliable information.

Stress levels can be influenced by the information which is communicated to passengers about an impending emergency and this information may originate from unintended as well as intended sources.   Information from cabin crew to passengers should always be unambiguous.   Ambiguous information tends to reinforce the existing levels of fear, and fearful passengers will thus interpret ambiguous messages in ways that will increase their fear.   For many aircraft passengers, the ambience of the cabin is in any case full of potentially ambiguous stimuli.   For example the noises of the engine, the changes in the configuration of the wings, the expressions on the faces of the cabin staff, all may be interpreted as evidence of impending disaster.   In these circumstances, ambiguous information from the cabin staff will not have the intended effect of reassurance but is more likely to increase alarm. As other passengers observe this alarm, their uncertainties concerning the threat are reinforced.

Airlines are, understandably, unenthusiastic to proclaim conspicuously the possibility of an emergency.   Passenger briefing is confined to the requirements of the regulations; the emergency exits are designed to blend inconspicuously with their surroundings; the signs labelling such exits are very low-key and discreet. Cabin staff are trained to encourage passengers to feel comfortable and relaxed. Such factors influence the approach to the provision of information about emergencies which is characterized by a reluctance to draw attention to the possibility of danger until there is no alternative.

This may be counterproductive in two ways. First, without prior warning, passengers are less likely to be prepared for the onset of the emergency. Secondly, it must be recognized that information derives not only from the cabin crew but also from events in the cabin and from other passengers. If there is a mismatch between the messages from the cabin crew and those from other more convincing sources, such as the presence of smoke, uncertainty will develop and the level of stress will rise.

The objective of communicating information about potential or actual emergencies is to ensure the achievement of optimal levels of arousal and to instil the belief that the actions to be taken will be effective either in preventing or in overcoming the threat.

Countermeasures to panic

It was noted above that panic may be triggered by the sudden loss of visual inputs. A Hardware solution is to ensure the provision of emergency lighting at night and of floor lighting in the presence of smoke. Advocates of smoke hoods have argued that the protection from smoke afforded by such hoods which allows vision to function under adverse conditions plays a major part in reducing the likelihood of panic.

Another trigger for panic is the perception that there is an insufficient means of escape. However, this perception is less likely to develop if there are pre-existing perceptions, inculcated by information and instruction, that the supply is adequate.

Panic in a large group may be triggered by the panic behaviour of a small number of individuals. To combat the spread of panic, it is important that flight attendants recognise the preliminary indications of dysfunctional behaviour and encourage appropriate behaviour patterns by means of assertive leadership and by removing influences which might quickly develop in undesirable directions.

It has been observed that those with a defined role are less liable to panic than those lacking such a role, particularly if the role involves caring for the welfare of others. Thus, the assignment of tasks to passengers, such as removing the hatch at the overwing exit, encourages the maintenance of effective behaviour. Extending this strategy to include as many passengers as possible in taking responsibility for a person or an activity is a potentially useful way to combat panic.

Strong, powerful leadership is likely to reduce the incidence of panic by providing firm direction and showing by example the appropriate behaviour to ensure survival. Good leadership can to some extent compensate for inadequate preparation where the tasks involved in escape are not unduly complex.

Since panic results in the disintegration of previously learned skills, it follows that prior rehearsal will act as a deterrent to the development of panic in cabin staff. Familiarity induced by frequent rehearsal of emergency procedures will reduce the risk of any breakdown in performance under stress. Such procedures should include not only the operation of Hardware devices but also the management of people behaving in unpredictable ways. It is doubtful whether some of the training schedules currently in use pay sufficient regard to this aspect of emergency procedures. Analogous levels of preparedness on the part of passengers could reduce the risk of inappropriate behaviour in the face of stress.

Countermeasure to freezing
Where freezing occurs, the most effective countermeasure is strong leadership whereby the individual is forcefully instructed to obey commands which require little cognitive effort for their execution. However, as freezing occurs typically in response to demands with which the individual does not know how to comply, it can best be avoided by adequate prior preparation.

Countermeasure to dependency
Those exhibiting dependent and helpless behaviour will follow strong leadership. Following the lead of others may often be adaptive. This is evidently so when cabin crew are active in managing the emergency situation. However, there are examples in accident reports of some passengers blindly following their fellows and perishing as a consequence when others, who took their fate into their own hands, were able to find safety.

CONCLUSIONS

There are two themes evident from this discussion of emergency management. The first is the importance of leadership in managing the appropriate level of arousal, in providing the amount and kind of information which will be functional for maximum safety, and in managing the behaviour of passengers when swift action must be taken. The second is the importance of prior preparation which, by resolving the uncertainty in a situation, serves to reduce disruptive levels of stress. The optimal form of advance preparation results in the ability to produce a well-rehearsed, smooth performance when an emergency takes place.

Both these aspects should be emphasized in the course of cabin

crew training. **Frequent rehearsal of emergency drills** including
realistic simulations of different types of passenger behaviour are
necessary in order that high levels of skilled performance may be
maintained.
  Whilst the value to passengers of the leadership provided by an
appropriately trained cabin crew is indisputable, it is impossible for
such personnel to respond to every passenger in all conceivable
emergencies. Cabin crew are few in number in relation to
passengers and may be incapacitated by an impact. Inevitably,
some passengers will have to depend on their own resources for
survival.
  There are two sources of evidence which lead to the conclusion
that a serious attempt to educate passengers in taking care of their
own safety is the optimal solution to the problem of passenger
survival. First, there are reports of survival of individuals who, for
reasons relating to their own personal history, made a habit before
a flight of reviewing the possibilities for escape should this become
necessary. When the necessity arose, they were in a position to
execute their plans, and often to help others (J13). Secondly,
accidents to aircraft carrying passengers who are themselves
professionally engaged in aviation, perhaps as flight crew or cabin
crew or in some other capacity, have shown survival rates which
would not be expected among typical passengers. On one occasion,
seagulls ingested into the engine of a McDonnell Douglas DC-10
caused the engine to disintegrate and the take-off was rejected.
Fire erupted on the right wing and 128 passengers evacuated in less
than a minute. There were no fatalities and only two serious
injuries. All these passengers were airline employees and all but
one had trained for an emergency (N10).
  Such considerations, together with the frequent violation of rules
noted in an earlier section above, lead to the conclusion that
improvements in cabin safety are dependent upon a more systematic
approach to the education of passengers. The dilemma of the dual
role of cabin staff has been discussed earlier (see p.61). The same
dilemma might be described alternatively as the dual role of
passengers: at one time customers expecting high standards of
comfort and service, at another victims of a disaster, struggling for
survival.
  There is a tendency for this duality in role perception to lead to
a dilution of the normal approach to the process of education and
training. Passengers exhibit little eagerness or even willingness to
learn. Attendants, aware of this apathy and frequently short of
time during the period between boarding and take-off, fulfil the
statutory obligation of providing a briefing, but frequently lack
conviction in the delivery.
  The mores of society encourage a lack of preparedness for
emergencies whether these exist in the air, in surface transport, in

the home, or elsewhere. It is for the industry along with legislative bodies to decide whether or not aviation is to lead the way by making serious attempts to increase the survival rate and reduce injury rates in the event of aircraft disasters. Should the response be positive, then there are changes to be made in the present way of utilizing system resources. Much of the Hardware is already in place. H-S and H-L interfaces create some problems as yet unresolved, and these have been discussed in previous sections. Many more significant difficulties arise concerning Software, most particularly in interfacing with human behaviour.

It is beyond doubt that safety would be enhanced were the passengers, as system users, more adequately prepared in order both to take precautionary steps to avoid the creation of hazards and to react appropriately during emergency conditions. In order to bring about substantial degrees of change, procedures would have to be developed to ensure that the education and training of passengers achieved standards far higher than those now prevailing.

The first step in the construction of any training programme comprises the compilation of a syllabus. The air transport industry should prepare and make quite explicit a syllabus for passengers which would be far more comprehensive than the present statutory briefings. Rather than being made up entirely of a list of "dos and don'ts", this should contain a reasonable amount of explanatory material. For example, it should be made clear that the smoking prohibition in the toilet compartment draws its authority from an issue of safety, not from the possibility of complaints about the odour of tobacco. Rules concerning Dangerous Goods on board, intoxication, and baggage restrictions should be included in the syllabus. The advantage of continually wearing a seat belt should be explained. Details concerning the possible emergencies, and schedules of associated equipment and procedures need to be fully comprehended.

Following the preparation of an agreed syllabus, the next stage involves decisions concerning the method of teaching. Booklets and video cassettes available for purchase at airports and elsewhere provide a possible solution. It has been suggested that passengers might be offered "hands-on" experience of safety equipment in mock-ups in order to add practical skills to their knowledge (B14). By such methods, passengers would have the opportunity to prepare themselves in advance of the flight. The short period available between boarding the aircraft and take-off could be used by cabin crew to draw attention to any special conditions, such as inoperative doors, and to remind each passenger of the location of exits and safety equipment in relation to the particular seat allocated.

Ideally, the final component of the safety training programme would comprise an assessment procedure, the successful completion of which would provide the passenger with the authority to fly.

The objections to such a systematic approach to passenger education are not difficult to predict. Time and cost play a substantial part. Most especially, the image of passenger aviation as a safe and enjoyable method of travel would not be enhanced by an emphasis upon preparation for disaster. Any one airline might well be reluctant to launch such a programme and risk losing customers to commercial competitors. Arguments concerning individual freedoms might be advanced. The long-term outlook is more difficult to assess. Lifeboat drill on passenger liners has long been established and accepted. The public might be inclined to revise current attitudes towards risks in the air, much as popular opinion has changed dramatically with respect to smoking, to some dietary habits, and to the need for environmental conservation.

REVIEW

The behaviour of people in emergency situations is not amenable to experimental investigation with the result that increased understanding must proceed relatively slowly. However, observations of genuine emergencies, combined with theoretical models of behavioural mechanisms, provide a basis for the formulation of recommendations concerning the management of emergencies. Greater awareness within the air transport industry of behavioural management would be advantageous.

It is well-established that passengers' lack of preparedness leads to increases in the toll of death and injury in the event of accidents. The only way to bring about substantial changes in this area is to introduce systematic programmes of education and training based upon a carefully constructed syllabus defining the necessary knowledge and skills which passengers require. The objections to such a programme are fairly obvious, and the current position may well be allowed to continue.

11   After the Emergency

Those who survive emergencies may find themselves to be suffering from the after-effects of stress. The support offered by families and friends does not always suffice. Prompt expert assistance is necessary to guide victims towards the restoration of a normal life.

It is a mistake to suppose that the end of an emergency situation spells the end of the trauma for the individuals involved, even where there is no physical injury. Disasters, both natural and man-made, which have occurred in recent years have shown that some individuals live in the shadow of the event sometimes for periods of several years afterwards and that skilled help is needed to assist them to come to terms with the horror of their experience and integrate it into their lives.

POST-TRAUMATIC STRESS DISORDER

That the disturbed emotional states of individuals who have survived disaster are becoming increasingly recognized is evidenced by the emergence of the label "post-traumatic stress disorder" (PTSD). This refers to the disturbances, both physical and mental, which may follow from the experience of events of a very frightening or horrific nature.   These disturbances include sleep loss, nightmares, digestive disorders, headaches;   depression, feelings of guilt, anger, shame, despair;   cognitive failures such as memory lapses and loss of concentration and regressive behaviours such as bed-wetting. There may be an apparently compulsive need to talk about the incident or there may be a blocking of the memories of the experience.
   In the immediate aftermath of the disaster, the survivors are the subject of concentrated attention from the media and from their family and friends. Offers of help tend to flow in. Special funds

may be organised to channel the expressions of sympathy felt by the general public into financial assistance for survivors.   The high levels of positive stimulation engendered in survivors during the initial period of survival, including their feelings of profound relief that their ordeal is over, may create the impression that they have "come through in good shape" and that, after a hot bath and a good meal, those who are uninjured will be ready to assume their normal roles in life.

It is later, when the public interest has waned and when family and friends believe that sufficient time has been allowed for recovery, that the experience of symptoms is likely to begin. However, this is when the least amount of sympathetic attention is likely to be paid to the sufferer, who may be thought to be malingering, and whose compulsive need to talk about the events becomes for the audience a tedious bore.

The development of PTSD

Whilst the experience of disaster may not necessarily include bereavement, the post-traumatic stress disorder shows certain similarities to the pathological grief reactions which sometimes occur in response to bereavement.

Grief is the natural reaction to loss and it functions to loosen bonds with the deceased person, to facilitate the readjustment of the bereaved individual to an environment from which the deceased person is absent, and to prepare the way for the formation of new relationships.   The normal grief response typically includes withdrawal, sleep disturbance, loss of appetite and periods of intense emotional distress.   In addition to these "acceptable" symptoms are those which the individual may find less acceptable, namely feelings of hostility towards the deceased person, guilt on account of this hostility, shame, and fear.

It is necessary to work through the grief if readjustment to the new situation is to take place.   The process of working through the grief involves accepting the feelings and experiencing their full impact;   expressing the feelings and allowing the intensity gradually to diminish.   However, this process may be hindered if the bereaved person tries to avoid the pain and distress associated with the feelings and expression of grief.   In addition, the negative feelings towards the deceased person may be so strong that they are obstacles to the successful working through of grief.   Instead, distorted and pathological grief reactions result which may include searching for the deceased person, hallucinations, depression, anxiety, feelings of worthlessness.   Feelings of aggression, in particular towards the deceased person, may be so difficult to accept that the resulting guilt feelings sometimes lead to suicide.

Like those who have been bereaved, those who have experienced a disaster have faced a major disruption of their lives, with attendant strong negative emotions of shock, fear, hostility, shame, and loss. The people, events, and objects, which before the disaster were neutral or positive in their emotional impact, become conditioned stimuli for powerful negative emotional responses by association with the trauma of the disaster. Because of the distress caused by these emotions, the traumatised individual wants above all to avoid them. Thus phobias develop. Both pathological grief and post-traumatic stress disorder have their roots in the avoidance of the pain and distress which is a necessary accompaniment for working through the grief.

There is also a cognitive dimension to PTSD. After experiencing an event so devastating but apparently arbitrary, individuals attempt to search for meaning, to construct a framework into which the experience may be fitted so that some order may be imposed on the confusion. This need to make sense of the environment can be satisfied in ways which are helpful to the individual or in ways which contribute to increased distress. In the latter case, the meaning given to the event may involve negative self-evaluation on the part of the individual (for example, personal inadequacy), the belief that others were to blame for the event, or the belief that the event was avoidable. All these attributions may serve to increase the stress.

The knowledge that the environment is predictable and controll-able, that certain actions will result in certain outcomes, is a key factor in the prevention of stress. However, the experience of a disaster may provide overpowering evidence that the environment is not controllable. Where the individual can perceive no connection between his actions and the outcome, then "learned helplessness" may result. This gives rise to feelings of futility, apathy and depression.

SECONDARY PROBLEMS

If those who have experienced a disaster are unable to tolerate the subsequent distress, they may seek refuge in chemical methods for the relief of their pain. Drugs (which may be medically prescribed) and alcohol very readily dull the unbearable anguish aroused by painful thoughts and memories but they are temporary solutions and leave untouched the underlying stress. The use of such solutions also prevents the development of effective coping behaviour. In addition, these crutches soon themselves become sources of difficul-ty, and problems arise which are the consequences of attempts to cope with the original stress. Alcoholism and drug dependency lead to a spiral of further problems, frequently involving estrangement from close family who become unable to tolerate the violence and alienation which frequently accompany substance abuse.

Those who have been involved in a disaster may feel that only those who have also undergone the experience of the disaster have the capacity to understand them and this may also lead to estrangement from close family who were not directly involved.   In this way, the individual is cut off from possible sources of support and the increasing social isolation is likely to exacerbate the symptoms already present.

Other consequential problems result when the familial role (bread-winner, carer) of the traumatised individual is assumed of necessity by another family member.   Attempts at readjustment to the roles of the pre-disaster situation may be unsuccessful, and thus lead to the break-up of the family.

INTERVENTION

It should be noted that stressed individuals do not uniformly exhibit heightened overt emotionality;   some manifest their distress by excessive withdrawal, giving the impression to the untutored eye that they are untouched by their experience.   This, in turn, may introduce additional problems.   Furthermore, there may be a tendency following the recognition of PTSD for the popular view to swing from the expectation that all survivors will rejoice at their escape to the expectation that all survivors will demonstrate severe stress symptoms.   While the proportion of those affected severely enough to require special help has been put as high as 80%, that leaves a significant proportion who are able to cope without it. Clearly, the issue of intervention must be treated with great sensitivity so that special help is targeted to those who need it.

For some people, the social support of family and friends may be sufficient.   Friends who are able to provide the appropriate listening skills and who understand the need for the repetitive ventilation of feelings of grief, anger, fear, and sometimes self-reproach, enable the necessary catharsis to take place.   However, not all close family and friends are able to cope with their own pain at the distress of the individual and, instead of listening, either respond with inappropriate expressions of comfort or attempts to divert attention to other matters.

Where special help is required, the role of institutional psychiatry should be minimized.   This is for a number of reasons.   There is still a tendency to stigmatise people who have received psychiatric treatment, and this in turn influences negatively the self-image of the individuals concerned.   Perhaps more importantly, referral to a psychiatric clinic is likely to involve a period of waiting when what is required is immediate help.

In recent years, there has been an increasing awareness among the general public of the after-effects of disaster.   Programmes on television and articles in newspapers have helped to extend an appreciation of the problems encountered by disaster victims. Both social service departments of government and voluntary groups such as The Samaritans have become more organized in their delivery of support to those who have been involved in major accidents. The need for urgency in the provision of services close in time and (where appropriate) location to the disaster has also been recognised. After the fire at King's Cross Underground Station in London in 1987, psychologists provided help to the most severely burned victims by the following day and a "Helpline" was established by means of which counselling could be offered to those who felt in need of help.

The most effective method of intervention is to provide a situation in which the stressed individuals can find it acceptable to talk about the disaster, to express their feelings of fear of being exposed again to a similar event, of grief at their loss, and their hostility and anger towards those believed to be responsible for the disaster. This scenario is likely to include others who have also experienced the disaster, and who can share fully in the feelings that are expressed. It may be helpful to teach relaxation techniques and to encourage physical exercise. In some cases where phobias have developed, behavioural techniques such as desensitization are effective.

AIRCRAFT DISASTERS

These are not different from other disasters in their impact on individuals. Like the professionals concerned in other public transport disasters, however, and as distinct from the travelling public, the aircraft crew will react in ways which reflect their professional involvement in the incident, and which may be more or less intense as a consequence.

The major concern of the cabin crew is safety and this is the focus of much of their training. They learn to recognize danger signals and to prepare for action in the event of emergencies. Drills are rehearsed so that they become familiar behaviour patterns which can be activated immediately the need arises. This training provides a certain amount of "stress inoculation", preparing the individual to know what to expect, and what to do in order to respond quickly and effectively to a disaster. All this acts to reduce any tendency to post-disaster stress.

However, if for any reason the cabin crew members are unable to carry out their duties (they may be injured, the exit for which they are responsible may be blocked and unusable), they may suffer intense feelings of guilt arising from failure to meet their own professional standards. When friends and colleagues perish in the disaster, they may suffer "survivor" guilt, and this is often compounded by anger towards those felt to be responsible. The grief response may be delayed until all the tasks associated with the incident are completed and help may be needed to "unfreeze" the emotions. Phobias concerning aircraft or aspects of flight may develop which could seriously interfere with the ability to continue working in the air.

Three factors are important in reducing the incidence and effects of post-traumatic stress disorders. These are the provision of help close in time and place to the onset of the stress, the expectation that the individuals will overcome their difficulties, and that they will be able to resume their ordinary lives. It would be advisable to have available an organizational structure standing in readiness to provide prompt specialized support for traumatized cabin attendants. To some extent, the Employee Assistance Program (EAP) established by the US Association of Flight Attendants might assume such a role. Essentially the EAP offers a service of mutual support by the members of the peer group which comprises this peripatetic work-force, supported by a sub-group selected for their special diagnostic and counselling skills.

Potential victims of PTSD are already identified by virtue of their presence during an incident and should have immediate expert attention available to them. The formal acceptance by the airline of the possibility of intense emotional reactions to traumatic events, together with a recognition of the temporary nature of the problem, would reduce the incidence of long-term post-disaster stress and help those affected to resume their normal lives.

12 Human Factors in
Design and Management

Human Factors has a significant part to play in both the design phase of a system and in the on-going management. In view of the frequent lack of adequate HF expertise, the audit serves as a practical starting point.

System design is the art of creating or procuring appropriate resources and integrating the component parts into an effective whole. System management is concerned to achieve safe and efficient performance within a changing environment.

Success in the processes of design and management is dependent upon a considerable amount of teamwork in all but the simplest of cases. Each of the basic resources - Hardware, Liveware, and Software - calls for its own expertise often further subdivided to provide authoritative knowledge and skills concerning the diverse facets within a single resource type.

A mere collection of experts does not comprise a team. Each person must have a working knowledge of the areas covered by his colleagues. Without a shared conceptual framework and a common language, communication is impossible. Problems must be identified, analyzed and solved by the team, often using iterative approaches to the final solution. In order to achieve the necessary levels of coordination, careful attention must be paid to the structure of the team and its method of working.

SYSTEM DESIGN

At an early stage in design, some basic HF data, which have long been familiar to aircraft designers, are used to convert payload or cabin floor area into a value expressing passenger capacity. This serves as an example of the convenience derived from knowledge spreading between disciplines in a design team. In an analogous way, HF personnel working in the aviation sector acquire some general knowledge concerning Hardware and Software, and are thus able to contribute economically to the design process.

Two groups of Liveware must be considered by the HF specialist concerned with the aircraft cabin.   The needs of passengers, the system users, will dictate recommendations in such areas as space utilization for sitting and for moving about, lighting, ventilation, noise.   The needs of cabin staff, the system operators, are largely associated with their work in providing a service to passengers. Both groups require the protection of emergency systems.

The HF contribution develops in parallel with other activities of the design team.   As the design of the cabin evolves, all the aspects of its eventual occupation and use by human beings must be evaluated.   Servicing, cleaning, and maintenance functions, in addi- tion to normal operation, require attention.   The location, availability, and detailed design of all portable items in the cabin will eventually become an object of study.

Software, in the form of airworthiness requirements, will have been influential throughout the design process. Similarly, environ- mental forces such as market requirements, costs, and timescales will have played their part in shaping decisions.

In conjunction with operational input, including that from potential customers, developmental work upon standard operating procedures - another type of Software - will begin.   It may be necessary for some iteration to take place should it appear that the provisional Hardware decisions are creating problems of an operational type.   In this way it eventually becomes possible for all the system interfaces to become harmonized, and for a complete specification of the tasks within the cabin to be prepared.

Throughout the design process, two major criteria will have guided the HF contribution.   First, the well-being of personnel requires promotion.   This notion includes not only the protection from traumatic accidental injuries, but also takes account of the long- term effects on human beings of such hazards as poor posture or exposure to noise.   The well-being both of system users and of operators is involved and proper compromises must be made in the event of potential incompatibilities.

The second criterion is that of system effectiveness; goals are to be achieved at minimum cost.   Interfaces are designed to promote ease and speed in the performance of tasks, and steps are taken to minimize sources of potential error.   A variety of job aids in the form of manuals, checklists, placards, conversion tables, etc. can contribute substantially to effective performance.

In view of the complexity of the design process, simulation may provide helpful ways of evaluating provisional decisions.   At its simplest level, simulation of system performance may be carried out using paper-and-pencil devices such as activity charts and time-line diagrams.   Physical aspects of cabin layout and the location of equipment may be examined using full-scale mock-ups of part or all of the cabin.   Well-built mock-ups may have later applications within the training programme.

As the design process nears completion, job specifications for the system operators can be finalized, and HF attention is directed towards the tasks of personnel selection and training.

SYSTEM MANAGEMENT

Within the lifecycle of an aircraft cabin there will be many changes in the deployment of resources. Equipment will be modified or replaced;   operating procedures will change;   individual personnel will vary and manning policies may be modified.

Due to the highly interactive nature of SHEL systems, a small change in one component may lead to disturbances elsewhere. These need to be anticipated if the system is to maintain its equilibrium.   A Hardware modification to an item of safety equipment, for example, may lead to the necessity for change in the procedure defining its use.   This, in turn, leads to the necessity to modify manuals, checklists and placards, and to amend the initial and recurrent training programmes.   In the event that some of these tasks are overlooked, then a hazard may be created.   An emergency may not be handled effectively;   injury may result from deficiencies in operation.   Alternatively, when mismatches come to the attention of system operators, there is a tendency for them to create their own solutions by "fixing" the Hardware or departing from the standard procedures.   The full ramifications of such unsystematic modifications are not always self-evident and unforeseen consequences may be dire.

Change is not confined to the system resources.   The economic, political and social Environment in which airlines operate is constantly subject to flux.   Prosperity, and even survival, depend upon the ability to adjust to prevailing circumstances.   Some aspects of social change take place gradually over long periods of time. Others occur more dramatically as the result of a specific event.   In either case the correct anticipation of change can be extremely valuable whether in the form of trend forecasting, or predicting the outcome of particular landmarks.

Many examples might be quoted to illustrate the ways in which the L-E interfaces within aviation have been subjected to change with the passage of time.   Relatively low-cost fares combined with increasing prosperity have brought about large increases in holiday traffic.   This has led to changes in the type of clientele carried with the consequent effects upon the behaviour of passengers and the type of service expected on board.   Such trends are more apparent in some countries than in others.

Policies relating to the recruitment and employment of women have, in many parts of the world, undergone considerable development since the first stewardesses were engaged in the 1930s. Many airlines now offer progressive career opportunities to female employees.

Events on the scale of world wars have brought civilian air transport to a virtual standstill. Much less traumatic, but still of considerable significance, have been political events such as the introduction of "deregulation", the removal of long-standing controls upon the conditions of commercial operation. This has led to intense competition between companies with the consequent changes in personal income, working conditions, and career prospects of cabin staff.

In order to cope successfully with all such aspects of the Environment, whether the changes are gradual or sudden, it is necessary for companies to be keenly aware of the significance of L-E interfaces within the SHEL system, and to introduce management policies which deploy system resources to the best advantage.

THE HUMAN FACTORS AUDIT

Were systems to be designed and managed at a level of perfection, the audit process would be unnecessary. In practice, such heights are seldom achieved. The somewhat idealized account of the role of HF in system design outlined above does not provide an accurate representation of every design process. Neither is every management team equipped to deal competently with all the Liveware problems within its ambit. Several explanations may be advanced to account for the neglect of Liveware issues. HF is a comparatively new technology and does not enjoy the established position of older disciplines. There is persistent adherence to the "HF is only common sense" view. The adaptability of Liveware components frequently provides temporary camouflage for the penalties resulting from unskilled interfacing.

The audit may be regarded as an error-detection and correction device applied to the processes of system design and management. The first phase comprises a review of all those parts of an organization in which Liveware components become involved whether as system operators or users. Errors, difficulties, delays, incidents, and near-misses are examined. Interfaces are checked in relation to the well-being of personnel and the effectiveness of system performance. Simulations may be employed to investigate the effect of errors in a variety of scenarios.

This descriptive information provides the basis of the second phase of the audit, the objective of which is to devise modifications to enhance the system. Such modification might be located at any point within the system from the recruitment, selection and training of personnel to adjustments in the system Hardware or Software. Clearly, any proposed changes would require evaluation in terms of cost-benefit estimation.

REVIEW

Human Factors can play a crucial role throughout the processes of design and management of the aircraft cabin. Generally, those interfaces which include a Liveware component tend to receive the least expert attention. This situation is not without irony, since human beings are the single resource having intrinsic value.

Appendices

APPENDIX 1 **List of Abbreviations**

AAIB	Air Accident Investigation Branch (UK)
AC	Advisory Circular
AFA	Association of Flight Attendants (US)
AMK	Anti-misting kerosene
AT&T	Aircraft Transport and Travel
ATC	Air Traffic Control
BCAR	British Civil Airworthiness Requirements
BOAC	British Overseas Airways Corporation
CAA	Civil Aviation Authority (UK)
CAB	Civil Aeronautics Board (US)
CAMI	Civil Aeromedical Institute
CAT	Clear Air Turbulence
CID	Controlled impact demonstration
CPR	Cardiopulmonary Resuscitation
dB	decibel(s)
dBA	decibel(s) A-scale
EAP	Employee Assistance Programme
ECU	Environmental Control Unit
ELT	Emergency Locatory Transmitter
ETS	Environmental Tobacco Smoke
FAA	Federal Aviation Administration
FAR	Federal Aviation Regulation
FSF	Flight Safety Foundation
ft	foot (feet)
G	Ratio between magnitude of an acceleration and that produced by Earth's gravity
GAS	General Adaptation Syndrome
HF	Human Factors
IATA	International Air Transport Association
ICAO	International Civil Aviation Organization
in	inch(es)
kg	kilogramme(s)
KLM	Koninklijke Luchtvaart Maatschappij (Royal Dutch Airlines)

l	litre(s)
lb	pound(s)
LOFT	Line Oriented Flight Training
m	metre(s)
mm	millimetre(s)
MEL	Minimum Equipment List
min	minute(s)
MPD	Maximum permissible dose
N	Newton(s)
NASA	National Aeronautics and Space Administration (US)
NPRM	Notice of Proposed Rule Making
NTSB	National Transportation Safety Board
PA	Public Address
ppmv	parts per million by volume
PTSD	Post-Traumatic Stress Disorder
RH	Relative Humidity
s	second(s)
SAE	Society of Automotive Engineers
SAFER	Special Aviation Fire & Explosion Reduction Advisory Committee
SAS	Scandinavian Airlines System
SI	Système Internationale d'Unités
SOP	Standard Operation Procedure
SPL	Sound Pressure Level
TUC	Time of Useful Consciousness
UK	United Kingdom
US	United States (of America)

APPENDIX 2   **Regulations governing the design and operation
                     of transport aircraft**

The International Civil Aviation Organization (ICAO), established
under the auspices of the United Nations in 1946, provides sets of
requirements relating to the design and operation of transport
aircraft. The Annexes to the Chicago Convention, ICAO's cha rter,
contain the technical material. The Annexes are listed in Table
A2.1. Although compliance is a voluntary matter, the requirements
are normally accepted by the signatories to the Convention and
incorporated into the legislative system of each state.
   In this way a high degree of international standardization is
achieved, and each state is encouraged to attain acceptable levels
of safety and efficiency.
   Within the Annexes, the requirements are stated in a very general
way, setting out the various facets to which attention should be
paid. Additional publications are produced which enlarge upon these
outlines. For example, one of the more recently adop ted Annexes,
No.18 "The Safe Transportation of Dangerous Goods by Air", is
supplemented by ICAO Document 9284, "Technical Instructions for
the Safe Transportation of Goods by Air". This latter volume
provides highly detailed guidance concerning each of nine different
categories of dangerous substances. It is a statutory requirement in
the UK that dangerous goods must be carried in accordance with
these ICAO Technical Instructions.
   In the present book, examples are quoted of regulations which
apply either in the UK or the US. In both countries, the current
rule-making bodies have evolved following complex changes in the
legislative organizations responsible for safe design and operation of
aircraft.
   In the UK, design standards were formerly within the province of
the Air Registration Board (ARB), an organization which was later
absorbed as a division of the Civil Aviation Authority (CAA).
Numerous defunct government ministries, such as the Ministry of
Aviation and the Board of Trade, have in the past held
responsibility for matters concerning aircraft operation.
   The primary statutory instrument is the Air Navigation Order, the
Schedules and Regulations of which contain the basic legislation
governing aircraft operation. These are published in CAP 393. The
CAA is the agency to which is delegated rule-making authority.
One division is concerned with the production of British Civil
Airworthiness Requirements (BCARs), the task once undertaken by
the ARB. Other branches of the CAA have responsibilities in such
areas as crew licencing and air traffic control. Cooperation with
other European states in specifying a code for aircraft design and
construction has led to the production of Joint Airworthiness
Requirements (JARs).

Public transport aircraft registered in the UK are required to hold an Air Operator's Certificate (AOC) granted by the CAA. The requirements to be met in respect of equipment, organization, staffing, training, and other operational matters are set out in CAP 360.

In the US, similarly, a number of now defunct organizations, including the Bureau of Commerce and the Civil Aeronautics Authority, previously held responsibility for aviation safety. In 1958, the Federal Aviation Agency was created. Eight years later, the organization became the Federal Aviation Administration (FAA) within the Department of Transportation. Following the creation of the Agency, the earlier regulations were recodified into the Federal Air Regulations (FAR) which are part of the Code of Federal Regulations (CFR). The FARs are divided into a number of sections. Part 25, for example, is concerned with airworthiness standards, whereas Part 121 relates to operating requirements. The FAA also publishes Airworthiness Directives (AD) and Advisory Circulars (AC).

In the UK and the US, accident investigation is conducted by bodies which are independent of the regulatory authorities. In the UK, investigations are conducted by the Air Accident Investigation Branch (AAIB) and in the US by the National Transportation Safety Board (NTSB).

Table A2.1   The Annexes to the Chicago Convention

1.	Personnel licencing
2.	Rules of the air
3.	Meteorological services for international air navigation
4.	Aeronautical charts
5.	Units of measurement to be used in air and ground operations
6.	Operation of aircraft
7.	Aircraft nationality and registration marks
8.	Airworthiness of aircraft
9.	Facilitation
10.	Aeronautical telecommunications
11.	Air traffic services
12.	Search and rescue
13.	Aircraft accident investigations
14.	Aerodromes
15.	Aeronautical information services
16.	Environmental protection
17.	Security
18.	The safe transport of dangerous goods by air

Table A2.2 shows some examples of the UK and US regulations in relation to the design and operation of the aircraft cabin.

Table A2.2 Some sources of regulation relating to the cabin

Topic	UK	US
Aisles	BCAR D4-3-4.2.5	FAR 25 815
Crew training	CAP 360 Pt1 Sect13-16	FAR 121.400 et seq.
Emergency exits	BCAR D4-3	FAR 25.807
Escape slides	BCAR D4-3-4.3	FAR 121.310
Evacuation demonstration	BCAR D4-3-4.5	FAR 25.803
Exit signs	BCAR D4-3-4.2.7	FAR 25.811
Fire extinguishers	BCAR D6-1	FAR 25.851
Life rafts	ANO Sched 5 Scales H,I,J,	FAR 121.339-340
Oxygen supply	ANO Sched 5 Scale K	FAR 25.1441
Seat belts & harnesses	BCAR D4-4-3	FAR 25.785
Seat strength	BCAR D3-8	FAR 25.561

Bibliography

1. ALPHABETIC LIST

A01. AIB Bulletin 12/79. (1979). London: Department of Trade, Accidents Investigation Branch.

A02. Air Accidents Investigation Branch. (1989). Report on the accident to Boeing 737-236 series 1, G-BGJL at Manchester International Airport on 22 August 1985. London: HMSO.

A03. Air Transport Users' Committee. (1988). Care in the Air. London: ATUC.

A04. Allen, R. (1978). KLM: A Pictorial History. Surrey: Ian Allan.

A05. Altmann, H.B., Johnson, D.A., & Blom, D.I. (1970). "Passenger emergency evacuation briefing cards: Recommendations for presentation style". In Proceedings of 8th Annual SAFE Symposium.

A06. American Medical Association Committee on Emergency Medical Services. (1982). "Medical aspects of transportation aboard commercial aircraft". Journal of the American Medical Association, 247(7), 1007-1011.

A07. Andrews, C. (1969). "Aircraft cabin fire protection". Flight Safety 3(1), 3-5.

A08. Anstey, J.B. (1969). "Escape slides for Concorde". Flight Safety 2(3), 14-16.

A09. Aircraft crash survival design guide. (1980). USARTL-TR-79-22, Applied Technology Laboratory. Fort Eustis, VA : US Army Research and Technology Laboratories, (AVRADCOM).

A10. Astrand, I. and Kilbom, A. (1969). "Physical demands on cabin personnel in civil aviation". Aerospace Medicine, 40, 885-890.

B01. Baker, G.W. & Chapman, D.W. (1962). Man and Society in Disaster. New York: Basic Books.

B02. Bailey, R.W. (1982). Human Performance Engineering. New York: Prentice Hall.

B03. Barthelmess, S. (1980). "Evacuation slides: history and new technology". Flight Safety Foundation, Cabin Crew Safety Bulletin, Nov/Dec.

B04.   Barthelmess, S. (1986). "Flight 797: A human-factors perspective, part 1". Flight Safety Focus, 86/7, 1-8.
B05.   Barthelmess, S. (1986). Flight 797: A human-factors perspective, part 2. Flight Safety Focus, 86/8, 1-9.
B06.   Beaty, D. (1976). The Water Jump. London: Secker and Warburg.
B07.   Beaty, D. (1986). The Complete Sky Traveller. London: Methuen.
B08.   Beckh, H.J. von. (1969). "Forward vs rearward facing passenger seats during emergency descent". Aerospace Medicine, 40(11), 1215-1218.
B09.   Beeding, E.L. & Mosely, J.P. (1960). Human Deceleration Tests, Airforce Missile Development Center, New Mexico, Rept No. AFMDC-TN-60-2.
B10.   Berkun, M.M., Bialek, H.M., Kern, R.P., & Yagi, K. (1962). Experimental studies of psychological stress in man. Psychology Monographs, 76(15), 1-39.
B11.   Birch, N. (1988). Passenger Protection Technology in Aircraft Accident Fires. Aldershot: Gower Technical Press.
B12.   Bird, J.A. (1968). "Training for safety". Flight Safety, 1(3), 10-14.
B13.   Blethrow, J.G., Garner, J.D., Lowrey, D.L., Busby, D.E., & Chandler, R.F. (1977). Emergency escape of handicapped air travellers. FAA-AM-77-11.
B14.   Blythe, N. (1986). "Passengers and professionals: the safety partnership". Paper presented by the Consumers' Association to the Royal Aeronautical Society's Symposium on Passenger Cabin Safety, 29 October.
B15.   Botteri, B.P., Gerstein, M., Horeff, T., & Parker, J. (1979). Aircraft Fire Safety. AGARD-AR-132. Neuilly sur Seine: NATO.
B16.   Brenneman, J.J. (1976). "The aircraft and fire from the operator's view". Paper presented to US National Fire Protection Association Seminar on Aircraft Rescue and Fire Fighting, Geneva.
B17.   Brenneman, J. (1985). "Inflight fire". In Proceedings of Cabin Safety Conference and Workshop, 11-14th Dec. 1984 DOT/ FAA/ASF100-85/01.
B18.   British Airways. Travelwise: Incapacitated Passengers. London: British Airways Customer Services.
B19.   British Airways. Travelwise: Cabin Baggage. London: British Airways Customer Services.
B20.   British Airways. Travelwise: Dangerous Articles in Baggage. London: British Airways Customer Services.
B21.   Brooks, B.M., Ruffell-Smith, H.P., & Ward, J.S. (1972). "An investigation of factors affecting the use of buses by both elderly and ambulant disabled persons". TRRL Contract No.

CON/3140/32.
B22. Brooks, P.W. (1961). The Modern Airliner: its origin and development. London: Putnam.
B23. Bruggink, G.M. (1983). "The uncontrollable cabin fire". International Journal of Aviation Safety, 1, 261-266.
B24. Bryan, M.E. (1976). "A tentative criterion for acceptable noise levels in passenger vehicles". Journal of Sound and Vibration, 525-535.
B25. Bryan, G.L. & Rigney, J.R. (1960). "Emotional behaviour of airline passengers". Flight Safety Foundation, Cabin Crew Safety Exchange.
B26. Buckle, P.W. & David, G.C. (1989). "Development of anthropometric selection criteria for airline cabin crew". Paper presented to Ergonomics Society Conference, April.
B27. Busby, D.E., Higgins, E.A., & Funkhauser, G.E. (1976). "Effects of physical activity of airline flight attendants on their time of useful consciousness in a rapid decompression". Aviation, Space and Environmental Medicine, 47(2), 117-120.
B28. Busby, D.E., Higgins, E.A., & Funkhauser, G.E. (1976). "Protection of airline flight attendants from hypoxia following rapid decompression". Aviation, Space and Environmental Medicine, 47(9), 942-944.
B29. Bellamy, J.E. (1989). Personal communication.
C01. Cameron, R.G. (1969). "Psychophysiological effects of flying on air hostesses". Aerospace Medicine, 40(9), 1018-1020.
C02. Campbell, E. (1986). "Passenger cabin smoke and fire". Flight Safety Focus, 86/2, 4-9.
C03. Canter, D. (Ed.) (1980). Fires and Human Behaviour. Chichester: Wiley.
C04. Chandler, R.E. (1985). "Seats, restraints and crash injury protection". In Proceedings of Cabin Safety Conference and Workshop, Dec. 11-14 1984. DOT/FAA/ASF/100 85/01.
C05. Chant, C. (1978). Aviation: an illustrated history. London: Book Club Associates.
C06. Chapman, P.J.C. & Chamberlain, D.A. (1987). "Death in the clouds". British Medical Journal, 294, 181.
C07. Chesterfield, B.P., Rasmussen, P.G. & Dillon, R.D. (1981). Emergency cabin lighting installation: an analysis of ceiling vs lower cabin-mounted lighting during evacuation trials. FAA-AM-81-7.
C08. Chisholm, D.M., Billings, C.A., & Bason, R. (1974). Behaviour of naive subjects during decompression: an evaluation of automatically presented passenger oxygen equipment. Aerospace Medicine, 45 (2), 123-127.
C09. Civil Aircraft Accident Report. (1981). No. 1/81. EW/C671. AIB, London.
C10. Civil Aviation Authority. (1982). Air Operators' Certificates -

Operation of Aircraft. CAP 360 Pt I.

C11. Civil Aviation Authority. (1982). Air Operators' Certificates - Arrangements for engineering support. CAP 360 Pt II.

C12. Civil Aviation Authority. (1982). The avoidance of excessive fatigue in aircrews. Guide to requirements.

C13. Civil Aviation Authority. (1985). Air Navigation: the Order and the Regulations. CAP 393.

C14. Civil Aviation Authority. (1986). Discussion Paper: Passenger smoke hoods. Paper No. S 836, Issue 1.

C15. Civil Aviation Authority. (1987). Airworthiness Notice No. 61. Issue 1, 16th March.

C16. Civil Aviation Authority. (1987).   Airworthiness Notice No. 79. Issue 2, 16th March.

C17. Civil Aviation Authority. (1988). Occurrence Digest Safety Matters, Aerospace, February.

C18. Civil Aviation Authority. (1982). Specification for operations manuals. CAP 450.

C19. Civil Aviation Safety Working Group. (1988). Cleared for takeoff?   London: TUC.

C20. Clegg, F. (1988). "Disasters: Can psychologists help survivors?" The Psychologist, 1(4), 134-135.

C21. Committee on Air Liner Cabin Air Quality. (1986). The Airliner Cabin Environment: Air Quality and Safety. Washington D.C.: National Academy Press.

C22. Combined analysis of workshop discussion. (1985). In Proceedings of Cabin Safety Conference and Workshop, Dec. 11-14 1984. DOT/FAA/ASF/100-85/01.

C23. Connell, F.G. (1976). "Training for the cabin emergency". Shell Aviation News, 433, 14-17.

C24. Cox, T. (1978). Stress. London: The Macmillan Press.

C25. Crawford, W.A. (1989). "Environmental Tobacco Smoke in Airliners - Health Issues". Aerospace, 16(7), 12-17 July.

C26. Cruickshank, J.M., Gorlin, R. & Jennett, B. (1988). "Air travel and thrombotic episodes: the economy class syndrome". The Lancet, August 27, 497-498.

C27   Cummins, RO. & Schubach, J. (1987). "Frequency and types of medical emergencies among commercial air travellers". Journal of the American Medical Association, 261(9), 1296-1299.

C28. Cuthbert, G. (1987). Flying to the Sun. London: Hodder and Stoughton.

D01. Daley, R. (1980). An American Saga: Juan Trippe and his Pan Am Empire. New York: Random House.

D02. Davison, G.C. & Neale, J.M. (1978). Abnormal Psychology: An experimental clinical approach. New York: John Wiley.

D03. Dean, R.D. & Whitaker, R.M. (1982). "Fear of flying". In Proceedings of the Human Factors Society 26th Annual

Meeting. 470-473.

D04.   Densley, F. (1970). "Overwater safety". Flight Safety, 3(4) 10-11.

D05.   Division of Labor Statistics and Research (1982). Work injuries and illnesses in California. San Francisco: Department of Industrial Relations.

D06.   Domogala, P. (1986). "Hijacking Part 1: The heritage". Flight Safety Focus, 86/3, 1-10.

D07.   Domogala, P. (1986). "Hijacking Part 2: Countermeasures in theory and practice". Flight Safety Focus, 86/4, 1-14.

D08.   Duffell, H.R.F. (1981). "Fire and smoke in aircraft fuselages". Flight Safety Focus, 1, 8-14.

D09.   Duffell, H.R.F. (1985). "An overview of aircraft occupant safety".   In Proceedings of Cabin Safety Conference and Workshop. Dec. 11-14. DOT/FAA/ASF100-85/01.

D10.   Dunn, B. (1981). "Safety awareness: A flight attendant's viewpoint".   Paper presented to Flight Safety Foundation's 34th International Air Safety Seminar, Acapulco.

D11.   Dunn, B. (1984). "Post-accident trauma and need for awareness".   Paper presented to First Anual Cabin Safety Symposium, University of Southern California, 7-9 February.

E01.   Edel, P.O., Carroll, J.J., Honaker, R.W., & Beckman, E.L. (1969). "Interval at sea level pressure required to prevent decompression sickness in humans who fly in commercial aircraft after diving". Aerospace Medicine, 40(10), 1105-1110.

E02.   Edwards, E. (1972). "Man and machine: systems for safety". Proceedings of BALPA Technical Symposium "Outlook on Safety", 21-36.

E03.   Edwards, E. (1988). "Introductory Overview". In E.L. Wiener and D.C. Nagel (Eds.), Human Factors in Aviation. San Diego: Academic Press.

E04.   Edwards, E. (1989). "Analysis of World Airline Accident Data". Unpublished report.

E05.   Edwards, M. (1976). Emergency exits of Public Service Vehicles.   Institute for Consumer Ergonomics, Loughborough University of Technology.

E06.   Edwards, M. (1977). "A design and evaluation study of handholds and footholds for emergency windows of Class III public service vehicles". Paper presented to Institute of Mechanical Engineers' Conference on Design, Construction and Operation of Public Service Vehicles, Cranfield Institute of Technology, July 11-13th.

E07.   Edwards, M. & Edwards, E. (1988).   "The ergonomics of the aircraft passenger cabin".   Proceedings of the Annual Conference of the Ergonomics Society, April.

E08.   Elder, P. (1985). "Airline seats: do they measure up to your needs?" Executive Travel, July, 18-22.

E09. Ellis, P. (1980). British Commercial Aircraft. London: Jane's Publishing Co.
E10. Endres, G.E. (1985). British Civil Aviation. Surrey: Ian Allan.
E11. Erneling, L. Orring, R. Joachimsson, A. (1988). Cabin Attendants' Working Environment Study. Stockholm: SAS
F01. Federal Aviation Administration. (1977). Advisory Circular: "Passenger safety information briefing and briefing cards". AC No. 121-24.
F02. Federal Aviation Administration. (1977). Advisory Circular: "Air Transportation of Mental Patients". AC No.120-34.
F03. Federal Aviation Administration. (1981). "Crewmember clothing: flammability standards". Withdrawal of notice of proposed rule making. Docket No. 14451.
F04. Federal Aviation Administration. (1983). "Floor proximity emergency escape path marking, Proposed Rules". Federal Register 48, 197.
F05. Federal Aviation Administration. (1984). Water Survival Staff Study. FAA-AM-633.
F06. Federal Aviation Administration. (1984). Air Carrier Operations Bulletin. 1-76-19.
F07. Federal Aviation Administration. (1984). Advisory Circular: "Passenger Information, FAR Part 135: Passenger Safety Information Briefing and Briefing Cards". AC No. 135-12.
F08. Federal Aviation Administration. (1986). "Improved seat safety standards, Proposed Rules". Federal Register 51, 137.
F09. Flight Safety Committee. (1985). "Inflight Fire". Flight Safety Focus, 85/4, 1-12.
F10. Flight Safety Committee. (1985). "B747 Rejected take-off accident". Flight Safety Focus, 85/5, 14-15.
F11. Flight Safety Committee. (1987). "B747 Evacuation". Flight Safety Focus, 87/8, 4-11.
F12. Flight Safety Committee. (1988). "Aloha Airlines B737 - preliminary". Flight Safety Focus, 88/5, 8-9.
F13. Flight Safety Foundation.(1969). "Cabin Crew Safety Exchange". Jan./Feb. 500-501.
F14. Flight Safety Foundation. (1970). "Survey of reactions of passengers to flight attendants' safety briefings". Cabin Crew Safety Exchange, 70, 501-502.
F15. Flight Safety Foundation. Safety Digest. (1987). 6, 10.
F16. Freedman, J.L., Birsky, J., & Cavoukian, A. (1980). "Environmental determinants of behaviour contagion: density and number". Basic and Applied Social Psychology 1(2), 155-161.
F17. Fritsch, O. & Santamaria, J. (1987). "Aviation safety - a review of the 1985 record". Flight Safety Focus, 87/1, 1-6.
F18. Fulton, H.B. (1985). "A Pilot's guide to cabin air quality and fire safety". New York State Journal of Medicine, July, 384-

388.

G01.   Gage, C.M. (1984). "Emergency evacuation problem areas".
       Paper presented to First Annual Cabin Safety Symposium,
       University of Southern California, 7-9 February.

G02.   Galer, I.A.R. (Ed.) (1987). Applied Ergonomics Handbook. 2nd
       Edition. London: Butterworth.

G03.   Gallup. (1984). The frequency of flying among the general
       public, 1984. Washington D.C.: US Air Transport Association.

G04.   Gaume, J.G. (1970). "Factors influencing the time of safe
       unconsciousness (TSU) for commercial jet passengers following
       cabin decompression". Aerospace Medicine, 41, 382-385.

G05.   Gaume, J.G. (1984). "Safety in the flight attendant's
       environment". Paper presented to the First Annual Cabin
       Safety Symposium, University of Southern California, 7-9
       February.

G06.   GP News. (1989). "BA improve emergency aid kit". General
       Practitioner, 19th May.

G07.   Gibbs-Smith, C.H. (1985). Aviation: an historical survey.
       London: HMSO.

G08.   Gibbons, H.L. (1984). "Inflight medical emergencies". Paper
       presented to the First Annual Cabin Safety Symposium,
       University of Southern California, 7-9 February.

G09.   Gilbert, G.C. (1976). "Cabin noise levels". Business and
       Commercial Aviation, July, 80-82, 84, 86.

G10.   Goldman, P. (1984). "Aircraft cabin safety in the United
       States: Fire is not the only hazard". International Journal of
       Aviation Safety, September, 185-186.

G11.   Goold, I. (1979). "Seats: the passenger/airline interface".
       Flight International, 15 September, 892-894.

G12.   Goold, I. (1980). "Safety and Comfort: the airliner cabin".
       Flight International, 16 February, 479-497.

G13.   Gould, R. (1989). "The ozone layer that nobody wants". The
       Guardian, 10 January.

G14.   Grant, W.M. (1974). Toxicology of the Eye. Springfield,
       Illinois: Charles C. Thomas.

G15.   Gray, J. (1971). The Psychology of Fear and Stress. London:
       Weidenfeld and Nicholson.

G16.   Green, R. (1969). "Cardiovascular disease in airline stewards".
       Aerospace Medicine, 40(11), 1264-1266.

G17.   Green, R. (1986). "Passenger behaviour in an emergency".
       Paper presented to the Royal Aeronautical Society's
       Symposium Cabin Safety, October 29.

G18.   Grieve, D. & Pheasant, S. (1982). "Biomechanics," in W.T.
       Singleton (Ed.) The Body at Work. Cambridge: CUP.

G19.   Guten, S. & Allen, V. (1972). "Likelihood of escape, likelihood
       of danger, and panic behaviour". Journal of Social
       Psychology, 87, 29-36.

G20. The Guardian. (1987). "Travellers put fire resistance top of safety league". 3rd November.

H01. Halberg, F. (1975). "The long-term effects of circadian rhythm manipulation - transmeridian dysynchronism fact or fancy". Paper presented to the BALPA Medical Symposium, London, October.

H02. Haward, L.R.C. (1969). "Passengers' attitude to fire risk". Flight Safety, 3(1), 8-9.

H03. Hawkins, F.H. (1980). "Sleep and body rhythm disturbance amongst flight crews in long range aviation: the problem and potential for relief". Unpublished M.Phil. Dissertation, University of Aston in Birmingham.

H04. Hawkins, F.H. (1987). "Human Factors in Flight". Aldershot: Gower Technical Press.

H05. Higgins, E.A. (1985). "Protective breathing: oxygen mask use/problems". In Proceedings of the Cabin Safety Conference and Workshop, December 11-14 1984. DOT/FAA/ASF 100 85/01.

H06. Hochschild, A.R. (1983). "The Managed Heart". California: University of California Press.

H07. Hockenberry, J. (1982). "A systems approach to long term task seating design". In Proceedings of the NATO Conference Series HF vol.16. Easterby R. et al (Eds.) Anthropometry and Biomechanics: Theory and Application. Plenum Press.

H08. Hoffler, G.W., Turner, H.S., Wick, R.L. & Billings, C.E. (1974). "Behaviour of naive subjects during rapid decompressions from 8000 to 30000ft". Aerospace Medicine, 5(2), 117-122.

H09. Holdener, F., Grab, F., & Joller-Jemelka, H. (1982). "Hepatitis virus in flying airline personnel". Aviation, Space and Environmental Medicine, 53, 587-590.

H10. Hopkin, H.A. (1986). "The market price for air safety". International Journal of Aviation Safety. (Reprinted in Crewman 1986, June/Sept.).

H11. Horne, J. (1988). Why we sleep: the function of sleep in humans and other mammals. Oxford: OUP.

H12. Horsfall, J. (1980). "Reducing fire hazards in modern aircraft cabins". Fire, November, 299-300.

H13. Horsfall, J. (1980). Cabin design for survival. Flight International, 28 June, 1493-1501.

I01. IES. (1981). IES Lighting Handbook (6th Edition). New York: Illuminating Engineering Society.

102. Iglesias, R., Terres, A., & Chavarria, A. (1980). "Disorders of menstrual cycle in airlines stewardesses". Aviation, Space and Environmental Medicine, 51, 518-520.

103. International Air Transport Association. (1968). Guidance material for passenger emergency evacuation briefing cards.

Montreal: IATA.

I04. International Air Transport Association. (1981). Incapacitated passengers travel guide. Canada: IATA.

I05. International Commission on Radiological Protection. (1966). "Radiation levels in SST operations". Health Physics, 12, 209.

I06. ICAO. (1974). "Accident to Varig Boeing 707 near Orly Airport, 11th July 1973". Aircraft Accident Digest No. 21

J01. Jacobson, I.D. & Martinez, J. (1974). "The comfort and satisfaction of air travellers. Basis for a descriptive model". Human Factors, 5, 46-55.

J02. Janis, I.L. (1962). "Psychological effects of warnings", in G.W. Baker & D.W. Chapman (Eds.) Man and Society in Disaster. New York: Basic Books.

J03. Jensen, R.S. (Ed.)(1989). Aviation Psychology. Aldershot: Gower Technical.

J04. Jerome, E.A. (1986). "The CAT hazard," Crewman, 13, 18-21.

J05. Johnson, D.A. (1971). "Behavioural inaction under stress conditions similar to the survivable aircraft accident". Paper presented to the 9th SAFE Symposium, Nevada.

J06. Johnson, D.A. (1973). "Effectiveness of video instructions on life jacket donning". Proceedings of the Human Factors Society Annual Meeting. Washington.

J07. Johnson, D.A. (1974). "Oxygen-use placard evaluation". Report No. MDC J6752. Douglas Aircraft Company, Long Beach, California.

J08. Johnson D.A. (1976). "Effectiveness of spoken instructions on passenger use of oxygen masks". Report No. MDC J7098. Douglas Aircraft Company, Long Beach, California.

J09. Johnson, D.A. (1978). "The effectiveness of sequential flashing lights to direct passengers to emergency exits". SAFE Journal, 8, 21-31.

J10. Johnson D.A. (1979). An investigation of factors affecting aircraft passenger attention to safety information presentations. FAA-IRC-79-1.

J11. Johnson D.A. (1980). "Emergency safety instructions: who attends and why?" Proceedings of International Conference on Ergonomics and Transport. Ergonomics Society: London.

J12. Johnson, D.A. (1983). "Passenger safety briefings". Flight Safety Foundation, Cabin Crew Safety Bulletin, 18(3) 1-4.

J13. Johnson, D.A. (1984). Just in Case. New York: Plenum Press.

J14. Johnson, D.A. (1984). "The effect of safety instructions on passenger panic and behavioural inaction following aircraft emergencies". Paper presented to First Annual Cabin Safety Symposium, University of Southern California, 7-9 February.

J15. Johnson, D.A., & Altman, H.B. (1972). "Analysis of passenger behaviour during four full-scale emergency evacuations". Report No. MDC J5423. Douglas Aircraft Co., Long Beach,

California.

K01. Kennedy, K.W. (1978). "Reach capability of men and women: a three-dimensional analysis". AMRL-TR-77-50. Dayton: Wright-Patterson AFB.

K02. Kilpatrick, M.A. & Brunstein, S.E. (1984). "Incident/accident survival and recovery". Paper presented to First Annual Cabin Safety Symposium, University of Southern California, 7-9 Feb.

K03. Knapp, S.C. and Knox, F.S. (1982). "Human response to fire". AGARD-LS-123.

K04. Koepp, S. (1987). "High anxiety and rage". Time, 20 July, 32-34.

K05. Koreltz, J.M. (1980). "Cabin Duties". Flight Safety Foundation, Cabin Crew Safety Bulletin. July/August.

K06. Kozlowski, L. (1976). "Safety on board. Evacuation procedures and training of cabin crew". Paper presented to US National Fire Protection Association Seminar on Aircraft Rescue and Fire Fighting, Geneva.

K07. Kraus, J.F. (1985). "Epidemiologic studies of health effects in commercial pilots and flight attendants". Journal of the Univeristy of Occupational and Environmental Health, Kitakyushu, Japan, 7, 32-44.

K08. Kroemer, K.H.E (1971). "Seating in plant and office". Journal of the American Industrial Hygiene Association, 32, 633-652.

K09. Kugihara, N., Misumi, J., Sato, S. & Shigeoko, K. (1982). "Experimental study of escape behaviour in a simulated panic situation. II Leadership in an emergency situation". Japanese Journal of Experimental Social Psychology, 21(2), 159-166.

L01. Lewis, M.F. (1970). "Vision through smoke hoods". in E.B. McFadden & R.C. Smith (Eds.) Protective Smoke Hood Studies. FAA AM-70-20.

L02. Luffsey, W.S. (1980). Testimony before US House of Representatives Subcommittee on Oversight and Review, "Cabin Safety: SAFER Committee Update - Aircraft passenger seat structural design" June 5.

L03. Lyman, J.L. & Mohler, S.R. (1985). "The airline passenger undergoing withdrawal or overdose from narcotics or other drugs". Aviation, Space and Environmental Medicine, 56, 451-456.

M01. Maclaren, R.B. (1986)."Cabin staff training". Paper presented to the Royal Aeronautical Society's Symposium on Passenger Cabin Safety, 29 October.

M02. Malmfors, T., Thorburn, D., & Westlin, A. (1989). "Air quality in passenger cabins of DC-9 and MD-80 aircraft". Environmental Technology Letters, 10, 613-628.

M03. Manley, R.J. (1984). "Aircraft cabin design for crash and fire considerations". Paper presented to First Annual Cabin Safety Symposium, University of Southern California, 7-9 February.

M04. Marshall, N. (1985). "Turbulence". In Proceedings of the Cabin Safety Conference and Workshop, Dec.11-14 1984. DOT/FAA/ASF100-85/01.

M05. Mason, C. (1975). "The problem passenger". PIA Air Safety, April, 12-15.

M06. McCormick, M.M. (1985). "Aircraft accident investigations: determining survival factors". Paper presented to First Annual Cabin Safety Symposium, University of Southern California, 7-9 February.

M07. McFadden, E.B. (1970). "Evacuation testing using dense theatrical smoke," in E.B. McFadden & R.C. Smith (Eds.) Protective Smoke Hood Studies. FAA AM-70-20.

M08. McFadden, E.B. & Swearingen, J.J. (1958). "Forces that may be exerted by man in operation of aircraft door handles". Human Factors, 1(1),16.

M09. McFadden, E.B., Swearingen, J.J., & Wheelwright, C.D. (1959). "The magnitude and direction of forces that man can exert in operating aircraft emergency exits". Human Factors, 1(4), 6.

M10. McFadden, E.B., Reynolds, H.I. & Funkhauser, G.E. (1967). A protective passenger smoke hood. FAA AM-67-4.

M11. McFadden,E.B. & Smith, R.C. (Eds.) (1970). Protective Smoke Hood Studies. FAA AM-70-20.

M12. McFadden, E.B. & Gibbons, H.L. (1970). "Smokehood effectiveness in a toxic environment". In E.B. McFadden and R.C. Smith (Eds.) Protective Smoke Hood Studies. FAA AM-70-20.

M13. McFarland, R.A. (1971). "Human factors in relation to the development of pressurized cabins". Aerospace Medicine, 12, 1303-1318.

M14. McKenzie, J.M., McFadden, E.B., Simpson, J.S. & Fowler, P.R. (1970). "Evaluation of leakage in protective smoke hoods".In E.B. McFadden and R.C. Smith (Eds.). Protective Smoke Hood Studies. FAA AM-70-20.

M15. McFarland, R.A. (1974). "Influence of changing time zones on air crews and passengers". Aerospace Medicine, 45(6), 648-658.

M16. McKie, L. (1987). "Slavery at 30,000 feet". The Guardian, 17 June.

M17. Meister, D. (1988). "References to reality in the practice of human factors". Human Factors Society Bulletin, 31(10),1-3.

M18. Melton, C.E. (1980). Effects of long-term exposure to low levels of ozone: a review. FAA AM-80-16.

M19. Millis, W.A. (1984). "The care of handicapped persons on scheduled flights". International Journal of Aviation Safety, December, 262-264.

M20. Mimpriss, J.G. (1980). "Fire and smoke in aircraft fuselages". Paper presented to Fire Safety Committee Discussion Group "Fires and Smoke in Aircraft Fuselages", London, December 17.

M21. "Models of perfection: the making of an air hostess". (1984). Business Traveller, Dec.

M22. Mohler, S.R., Nicogossian, A. & Margolies, R.A. (1980). "Emergency medicine and the airline passenger". Aviation, Space and Environmental Medicine, 51, 918-922.

M23. Mohler, S.R. (1976). "Idealised inflight airline medical kit: a committee report". Aviation, Space and Environmental Medicine, 47, 1094-1095

M24. Morrison, D., Hartley, L. & Kemp, D. (Eds.) (1986). Proceedings of the 23rd Annual Conference ESANZ, November 24-28.

M25. Moseley, H.G., Townsend, F.M., & Stembridge, V.A. (1958). "Prevention of death and injury in aircraft accidents". American Medical Association Archives of Industrial Health, 17, 111-117.

M26. Mouden, H. (1981). "Safety management". Paper presented to the Norwegian Society of Chartered Engineers, Trondheim, 8-10 January.

M27. Mouden, H. (1987). "Passenger protection and safety". Flight Safety Foundation, Cabin Crew Safety Bulletin, 22(4), 1-6.

M28. Mott, D. (1975). "Passenger escape from commercial aircraft". Paper presented to 27th International Seminar of the Flight Safety Foundation.

M29. Mott,D. (1984). "Cabin, cockpit and management coordination". Paper presented to First Annual Cabin Safety Symposium, University of Southern California, 7-9 February.

M30. Muir, H. and Marrison, C. (1989), Human factors in cabin safety. Aerospace, 16(4), 18-21.

M30A. Muir, H. Marrison, C. & Evans (1989). Aircraft evacuations: the effect of passenger motivation and cabin configuration adjacent to the exit. Paper 89019, London: CAA

M31. Munson, K. (1967). Civil Aircraft of Yesteryear. Surrey: Ian Allan.

M32. Murphy, D. (1984). "Interior design aspects". Paper presented to First Cabin Safety Symposium, University of Southern California, 7-9 February.

M33. Musbach, A. & Davis, B. (1980). Flight Attendant. New York: Crown Publishers Inc.

N01. Nagel, D.C. (1988). "Human error in aviation operations". In E.L. Wiener & D.C. Nagel (Eds.), Human Factors in Aviation. San Diego: Academic Press.

N02. NASA. (1978). Anthropometric Source Book. Reference Publication 1024. Washington D.C.: National Aeronautics and

Space Administration.

N03. National Research Council Committee on Airliner Cabin Air Quality. (1986). The Airliner Cabin Environment. Washington D.C.: National Academy Press.

N04. National Transportation Safety Board. (1972). Special Study. Passenger survival in turbojet ditchings (*A critical case review*). AAS-72-2.

N05. National Transportation Safety Board. (1972). Aircraft Accident Report. Boeing 747, 30 July 1971, San Francisco. AAR-72-17.

N06. National Transportation Safety Board. (1973). Special Study. In-flight safety of passengers and flight attendants aboard air carrier aircraft. AAS-73-1.

N07. National Transportation Safety Board. (1974). Special Study. Safety aspects of emergency evacuation from aircraft. AAS-74-3.

N08. National Transportation Safety Board. (1976). Aircraft Accident Report. Near mid-air collision between DC-10 and L-1011, 26th November, 1975. AAR-76-3.

N09. National Transportation Safety Board. (1976). Special Study. Chemically generated supplemental oxygen systems in DC-10 and L-1011 aircraft. AAS-76-1.

N10. National Transportation Safety Board. (1976). Aircraft Accident Report, DC-10, Jamaica, 12th November, 1975. AAR-76-19.

N11. National Transportation Safety Board. (1977). Special Study: US Air Carrier Accidents involving Fire 1965 through 1974, and Factors affecting the Statistics. AAS-77-1.

N12. National Transportation Safety Board. (1977). Aircraft Accident Report. Boeing 707, Pago Pago, 30th January 1974. AAR-77-7.

N13. National Transportation Safety Board. (1978). Aircraft Accident Report. Boeing 727, Florida, 8th May 1978. AAR-78-13.

N14. National Transportation Safety Board. (1978). Aircraft Accident Report. DC-9, Philadelphia, 23rd June 1976. AAR-78-2.

N15. National Transportation Safety Board. (1979). Aircraft Accident Report, DC-10, Los Angeles, 1 March 1978. AAR-79-1.

N16. National Transportation Safety Board. (1981). Special investigation report. SIR 81-4.

N17. National Transportation Safety Board. (1981). Cabin safety in large transport aircraft. AAS-81-2.

N18. National Transportation Safety Board. (1982). Aircraft Accident Report. Boeing 737, Washington, 13th January. AAR-82-8.

N19. National Transportation Safety Board. (1982). Aircraft Accident report. DC-10, Boston Harbour, 23rd January 1982. AAR-82-15.

N20. National Transportation Safety Board. (1984). Aircraft Accident Report. DC-9, Cincinatti, 2nd June 1983. AAR-84-9.

N21. National Transportation Safety Board. (1984). Aircraft Accident Report. L-1011, Miami, 5th May 1983. AAR-84-04.

N22. National Transportation Safety Board. (1985). Safety Study. Airline passenger safety education: a review of methods used to present safety information. SS-85-09.

N23. National Transportation Safety Board. (1985). Safety Study. Air carrier overwater emergency equipment and procedures. SS-85-102.

N24. National Transportation Safety Board. (1986). Aircraft Accident Report, L-1011, New York, 15th February. AAR-86.

N25. Newsom, W.A., Tredici, T.J., & Noble, L.E. (1968). "Danger of contact lenses at altitude". In Proceedings of 17th International Congress of Aviation and Space Medicine, Aug.5-8.

N26. Newsweek. (1984). "Can we keep the skies safe?" Jan. 30.

N27. Nicholson, A.S. & Ridd, J.E. (Eds.) (1988). Health, Safety and Ergonomics. London: Butterworth.

N28. Ninnescah. (1985). Survey of wheelchair passengers. 3(3), 3-5.

N29. Ninnescah. (1986). Blind passengers and air travel. 4(2), 6-10.

N30. Ninnescah. (1988). Security and disabled passengers. 6(2), 2-12.

O01. Oborne, D.J., & Clark, M.J. (1975). "Questionnaire surveys of passenger comfort". Applied Ergonomics, 6(2), 97-103.

O02. Owens, C.A. (1982). Flight Operations: a study of Flight Deck Management. London: Granada.

P01. Parker, J.F. & West, V.R. (Eds.) (1973). Bioastronautics Data Book. NASA SP-3006. Washington: National Aeronautics and Space Administration.

P02. Perry, I. (1979). "Survival". Flight International, 12 May, 1598.

P03. Pollard, D.W. (1979). "Injuries in air transport emergency evacuations". Aviation, Space and Environmental Medicine, 50, 943-947.

P04. Pollard, D.W., Steen, J.A., Biron, W.A., & Cremer, R.L. (1984). Cabin safety subject index. FAA-AM-84-1.

P05. Presidency of Civil Aviation Saudi Arabia. (1982). Aircraft Accident Report, Lockheed L-1011 HZ-AHK, 19 August 1980.

P06. Presidency of Civil Aviation Saudi Arabia. (1981). Report of accident to Lockheed L-1011, HZ-AHJ, 22 December 1980.

P07. Preston, F. (1978). "The health of female air cabin crew". Journal of Occupational Medicine, 20, 597-600.

P08. Preston, F. (1979). "Aircrew stress". Paper presented to the Dutch Airline Pilots' Association Symposium "Safety and

Efficiency: the next 50 years".

P09.   Preston, F.C. (1987). "Death in the clouds". British Medical Journal, 294, 374.

P10.   Preston, F., Ruffell Smith, H.P., & Sutton-Mattocks, V.M. (1973). "Sleep loss in air cabin crew". Aerospace Medicine, 44, 931-935.

P11.   Pronko, N.H. & Leith, W.R. (1956). "Behaviour under stress: a study of its disintegration". Psychological Reports, 2, 205-222.

Q01.   Quarantelli, E. (1954). "The nature and conditions of panic". American Journal of Sociology, 60, 267-275.

R01.   Ramsden, J.M. (1976). The Safe Airline. London: MacDonald & James.

R02.   Ramsden, J.M. (1983). "The survivable aircraft fire". Flight International, 13 August, 432-434.

R03.   Ramsden, J.M. (1986). "Safety standards under fire". Flight International, 25 October, 24-27.

R04.   Randall, L. (1979). A definitive study of your future as an airline steward/dess. New York: Richards Rosen Press Inc.

R05.   Randle, I.P.M., Turner, J.P., Hawkins, L.H. & Stubbs, D.A. (1986). Physiological responses and escape time during aircraft evacuation. In D. Morrison, L. Hartley & D. Kemp (Eds.), Proceedings of the 23rd Annual Conference ESANZ, 24-28 November.

R06.   Rasmussen, P.G., Garner, J.D., Blethrow, J.G., & Lowrey, D.L. (1979). Readability of self-illuminated signs in a smoke-obscured environment. FAA-AM-79-22.

R07.   Rasmussen, P.G., Chesterfield, B.P., & Lowrey, D.L. (1980). Readability of self-illuminated signs obscured by black fuel-fire smoke. FAA AM-80-13.

R08.   Rasmussen, P.G., & Chittum, C.G. (1986). The effect of proximity seating configuration on door removal time and flow rates through a Type III emergency exit. FAA-AAM-119-86-8.

R09.   Reason, J. (1974). Man in Motion: The Psychology of Travel. London: Weidenfeld and Nicholson.

R10.   Reed, D., Glaser, S., & Kaldor, J. (1980). "Ozone toxicity symptoms among flight attendants". American Journal of Industrial Medicine, 1, 43-54.

R11.   Rigby, L.V. (1968). "A predictive scale of aircraft emergencies". Human Factors, 10(5), 475.

R12.   Roebuck, J.A. & Levedahl, B.H. (1961). "Aircraft ground emergency exit design considerations". Human Factors, 3, 3, 174.

R13.   Rolfe, J.M. & Staples, K.J. (Eds.) (1987). Flight Simulation. Cambridge: CUP.

R14.   Ross, H.E. (1974). Behaviour and Perceptions in Strange

Environments.   London: George Allen and Unwin Ltd.

S01.   SAFER. (1980). Special Aviation Fire and Explosion Reduction Advisory Committee Final Report. FAA-ASF-80-4.

S02.   Sako, H. & Misumi, J. (1982). "An experimental study of the effects of the perceived likelihood of successful escape behaviour in a simulated panic situation". Japanese Journal of Experimental Social Psychology, 21(2), 141-148.

S03.   Salvendy, G. (Ed.) (1987).   Handbook of Human Factors. New York: Wiley.

S04.   Sanders, M.S. & McCormick, E.J. (1987).   Human Factors in Engineering and Design. New York: McGraw-Hill.

S05.   Sandland, A.J. (1969). "Fire fighting equipment for aircraft crash fires". Flight Safety, 3(1), 6-7.

S06.   Sarkos, C.P., Hill. R.G., & Howell, W.D. (1982). "The development and application of a fullscale widebody test article to study the behaviour of interior materials during a postcrash fuel fire". AGARD-LS-123.

S07.   Sarvesvaran, R. (1986). "Sudden natural deaths associated with commercial air travel". Med Sci Law, 26: 35-38.

S08.   Scheichl, L. (1977). "Survival after a crash".   Airport Forum, 77, 5, 71-87.

S09.   Schmidt, J.K. and Kysor K.P. (1987). Designing airlie passenger safety cards. In Proceedings of Human Factors Society Annual Meeting 1, 51-55.

S10.   Segal, J. & Luce, G.C. (1972). Sleep. New York: Coward-McCan.

S11.   Selye, H. (1956). The Stress of Life. New York: McGraw-Hill.

S12.   Sime, J.D. (1980). "The concept of panic". In D. Canter (Ed.), Fire and Human Behaviour. Chichester: Wiley.

S13.   Simonson, M.D. (1984). "Problem areas in flight attendants' health". Paper presented to the First Annual International Aircraft Cabin Safety Symposium.

S14.   Sinaiko, H.W., Guthrie, G.M., and Abbott, P.S. (1969). "Operating and maintaining complex military equipment: A study of training problems in the Republic of Vietnam". Institute for Defence Analysis Report P-501.

S15.   Singleton, W.T. (Ed.) (1982). The Body at Work: Biological Ergonomics. Cambridge: CUP.

S16.   Singleton, W.T. (1989). The Mind at Work. Cambridge: CUP.

S17.   Smith, R.C. (1970). "Effects of variations in safety briefings upon the use of protective smoke hoods". In E.B. McFadden and R.C. Smith (Eds.), Protective Smoke Hood Studies. FAA AM-70-20.

S18.   Snow, C.C., Carroll, J.J., & Allgood, M.A. (1970). Survival in emergency escape from passenger aircraft. FAA-AM-70-16.

S19.   Snyder, R.G. (1973). "Impact," in Parker, J.F. & West, V.R.

(Eds.). Bioastronautics Data Book, Washington: NASA.

S20. Snyder, R.G. (1977). "Advanced techniques in crash impact protection and emergency egress from air transport aircraft". AGARD-AG-221.

S21. Snyder, R.G. (1982). "Impact protection in air transport passenger seat design". Paper presented to SAE Congress, Anaheim.

S22. Society of Automotive Engineers. (1981). "Emergency placarding - internal and external". ARP 577B. Warrendale: SAE.

S23. Society of Automotive Engineers. (1983). "Passenger safety information cards". ARP 1384A. Warrendale: SAE.

S24. Sprogis, H. (1984). "Cockpit and cabin coordination". Paper presented to the First Annual Cabin Safety Symposium, University of Southern California, 7-9 February.

S25. Stapp, J.P. (1951). "Human exposure to linear deceleration; Part II The forward facing position and the development of a crash harness". A.F. Technical Report no. 5915.

S26. Stapp, J.P. (1970). "Voluntary human tolerance levels". In Gurdgian, Lange, Patrick and Thomas, (Eds.), Impact Injury and Crash Protection. Charles Thomas: Springfield, Illinois.

S27. Stapp. J.P. (1971). "Biodynamics of deceleration, impact and blast". In Rondel (Ed.), Aerospace Medicine. Baltimore: Williams and Wilkins.

S28. Stauffer W.A. & Wittlin, G. (1985). "Airframe structural integrity, design criteria". Proceedings of the Cabin Safety Conference and Workshop. Dec 11-14.1984 DOT/FAA/ASF/ 100-85/01

S29. Steiguer, D. de, & Salvidar, J.T. (1983). An analysis of potential protective breathing devices intended for use by aircraft passengers. FAA-AM-83-10.

S30. Stephens, D.G. (1979). "Developments in ride quality criteria". Noise Control Engineering, Jan/Feb, 6-14.

S31. St. John Turner, P. (1969). The Vickers Vimy. London: Patrick Stephens.

S32. Stroud, J. (1971). The World's Airliners. London Bodley Head.

S33. Swanson, C.E. (1970). "Anxiety and the consumer: fears of air travel". Flight Safety, 4(1), 15-17.

T01. Taylor, A.F. (1976). "An evaluation of world-wide transport aircraft fire experiences". Paper presented to the U.S. National Fire Protection Association's Seminar on Aircraft Rescue and Fire Fighting, Geneva.

T02. Taylor, A.F. (1986). "Accident statistics". Paper presented to the Royal Aeronautical Society's Symposium on Passenger Cabin Safety, October 29.

T03. Taylor, J.W.R. (1984). The Story of Flight. London: Chancellor Press.

T04. Technical   Group   Report.   (1968).   "Evacuation   and
      crashworthiness development program".   Washington DC:
      Aerospace Industries Association of America.

T05. Thompson, L.J. (1987). "Who should treat medical emergencies
      on commercial airliners?"   Aviation Medicine Quarterly, 1,
      125-129.

T06. Tobias, J.V. (1970). "Smoke hood tests: acoustic attenuation".
      In E.B. McFadden & R.C. Smith (Eds.), Protective Smoke
      Hood Studies. FAA AM-70-20.

T07. Turnbow, J.W., Carroll, D.F., Haley, J.L., & Robertson, J.H.
      (1969). "Crash survival design guide".   USAAVLABS
      Tech.Rept. 70-22.

U01. United States Air Force Design Handbook. (1974). Washington:
      Air Force Systems Command and National Aeronautics and
      Space Administration.

U02. United States Department of Transport. (1987). "Airline Cabin
      Air Quality: Report to Congress".

V01. Vant, J.H.B. (1986). "Aspects of health and safety in the
      passenger cabin".   Paper presented to the Royal Aeronautical
      Society's Symposium on Passenger Cabin Safety, October 29.

V02. Vestrey (Lord) & Kolesnik, E.G. (1982). Airship Saga.
      Dorset: Blandford Press.

V03. Vroom, V.H. & Deci, E.L. (Eds.), (1978). Management and
      Motivation. Harmondsworth: Penguin Books Ltd.

W01. Warren, D.V., (1986).   "CAA airworthiness requirements".
      Paper presented to the Royal Aeronautical Society's
      Symposium on Passenger Cabin Safety, October 29.

W02. Wakefield, R. (1986). "Death in the clouds". British Medical
      Journal, 293, 1642.

W03. Wiener, E.L. and Nagel, D.C. (Eds.) (1988). Human Factors in
      Aviation. San Diego: Academic Press.

W04. Wilkes, P. (1979). History of Aircraft. Bourne End: Spurbook.

W05. Wilson, K.G. & Helmholtz, H.F. (1947). Safety advantage of
      rearward seating in passenger aircraft. USAF Air Transport
      Command.

W06. Winkel, J. (1983). "On the manual handling of wide-body carts
      used by cabin attendants in civil aircraft".   Applied
      Ergonomics, 14(3), 162-168.

W07. Withey, W.R. (1982). "The provision of energy". In W.T.
      Singleton (Ed.), The Body at Work: Biological Ergonomics.
      Cambridge: CUP.

W08. Wolf, C. (1972). Aerotitis in air travel. Calif. Med.117: 10-12.

W09. Woodcock, A. & Davis, M. (1980). Catastrophe Theory.
      London: Penguin Books.

W10. Wright, P. (1988). "Functional literacy: reading and writing at
      work". Ergonomics, 31(3), 265-290.

Y01. Yaffe, M. (1987). Taking the Fear out of Flying. Newton

Abbot: David and Charles.

Y02.   Yates, A.D. (1982). "Cockpit/cabin communications".   Cabin
       Crew Safety Bulletin, Flight Safety Foundation, Jan/Feb.

Y03.   Yerkes, R.M. &   Dodson, J.D. (1908). "The relation of
       strength of stimulus to rapidity of habit forming".   Journal of
       Comparative and Neurological Psychology, 18, 459-482.

Y04.   Yost, C.A. & Oates, R.W. (1968).   Human survival in aircraft
       emergencies. NASA-CR-1262.

Z01.   Zurk, R. van, &   Toeset, J. (1984). "A new cabin emergency
       trainer".   International Journal of Aviation Safety, December,
       252-254.

2. SUBJECT CLASSIFICATION

Acceleration
    B09 N17 P01 S19 S20 S21 S25 S26 S27 U01
Accidents
    A01 A02 C09 F10 F12 F17 I06 M06 M25 N05 N08 N10 N12-
    N21 N24 P05 P06 S08 T02
Anthropometry & Seating
    B08 B26 C04 C16 C26 E07 E08 F08 G11 G18 H07 K01 K08
    N02 U01 W05
Aviation - General & Historical
    A04 B06 B07 B22 C05 C28 D01 E09 E10 GO7 M31 O02 S31
    S32 T03 V02 W04
Behaviour under stress
    B10 B25 C03 C20 C24 D03 D11 G15 G17 G19 J05 J14 J15
    K04 K09 P08 Q01 S02 S11 S12 S33 Y01 Y03
Cabin atmosphere
    C21 C25 E11 F18 I05 M02 M18 R10 U02
Cabin design
    B24 C04 C07 C16 E11 F04 H13 M03 M28 M32 O01 S21 S30
    U01 W06
Communication
    C18 F14 J02 J06 J07 J08 J10 J11 J12 J14 M29 N22 S17 S22
    W10 Y02
Crashworthiness
    A09 B08 C04 H13 M03 M06 M25 S19 S20 S21 S28 T04 T07
Decompression
    B27 B28 C08 E01 F12 G04 H05 H08 M13 N09 P06
Disability
    A06 B13 B18 B21 F02 F11 I04 M19 N28 N29 N30
Ditching
    D04 F05 J06 N04 N18 N23
Emergency exits
    A08 B03 B13 E05 E06 F04 F11 G01 J09 K06 M08 M09 M30
    N07 P03 R05 R08 R12 S18
Fatigue & sleep
    C12 H01 H03 H11 M15 P10 S10
Fire & Smoke
    A07 B11 B15 B16 B17 B23 C02 C03 C14 D08 F03 F09 H02
    H05 H12 K03 L01 M07 M10 M11 M12 M14 M20 N11 R02 R03
    R06 R07 S01 S05 S06 S17 S29 T01 T06
Health & medicine
    A06 A10 B18 C01 C06 C26 C28 E01 E11 G06 G08 G14 G16
    H09 I02 K07 L03 M18 M22 M23 N25 N27 P03 P07 P09 R10
    S07 S13 T05 V01 W02 W08
Human Factors (General)
    B02 B04 B05 E02 E03 G02 H04 J03 M17 N27 P01 R09 S03

S04 S15 S16 W03
Human Performance
    A10 B10 B13 B21 B27 C01 C08 D02 F16 H08 J05 J06 J07
    J09 J14 J15 K09 L01 M07 M08 M09 M11 M30 N01 P11 R05
    R06 R07 R08 R12 R14 S02 S19 W03 W06 W07 Y03
Lighting & Noise
    B24 C07 F04 G09 I01 J09 R06 R07 T06
Passenger briefing
    A05 B18 B19 B20 F01 F07 I03 I04 J10 J11 J12 J14 N22 S09
    S22 S23
Safety (General)
    A03 B01 B14 C19 C22 D09 D10 G05 G10 G12 H10 M26 M27
    N06 N17 P02 R01 R14 V01 W01
Selection & Training
    B12 B26 C23 K05 K06 M01 R04 R13 S14 Z01
Statistical data
    D05 E04 F17 G03 T02

Index